T0276066

SpringerBriefs in Astronomy

Series Editors
Martin Ratcliffe
Valley Center, Kansas, USA

Wolfgang Hillebrandt
MPI für Astrophysik, Garching, Germany

Michael Inglis
Long Island, New York, USA

David Weintraub
Vanderbilt University, Nashville, Tennessee, USA

More information about this series at http://www.springer.com/series/10090

Thomas G. Bisbas

The Interstellar Medium, Expanding Nebulae and Triggered Star Formation

Theory and Simulations

 Springer

Thomas G. Bisbas
Department of Physics and Astronomy
University College London
London, UK

ISSN 2191-9100 ISSN 2191-9119 (electronic)
SpringerBriefs in Astronomy
ISBN 978-3-319-26140-9 ISBN 978-3-319-26142-3 (eBook)
DOI 10.1007/978-3-319-26142-3

Library of Congress Control Number: 2015955145

Springer Cham Heidelberg New York Dordrecht London

Printed on acid-free paper

Springer International Publishing AG Switzerland is part of Springer Science+Business Media (www.springer.com)

To my parents Γιωργο and Ξανθη, my sister Αννα and my brother Ηλιας, and my nephew Σωτηρης with my best wishes

Preface

The book you hold serves as a handbook useful for graduate students and astronomers in their early career who wish to have a quick introduction to the concepts of the interstellar medium, nebulae, and triggered star formation. The present work is neither a textbook nor an overview. The aim is to provide some key work for the reader with the necessary updates in the field. At the same time particular effort has been paid toward an outreach type of presentation for the majority of the topics covered. Following that way of thinking and in order to bring the reader a step closer to the subject of nebulae, images taken from one of the world's most talented astronomy photographers, Kallias Ioannidis, have been used. The photographs you see in the book consist a fraction of the astounding gallery that Kallias Ioannidis holds. All of the images presented have been taken from various parts in Greece including Grevena, Thessaloníki, Kavala, and the beautiful island of Thassos.

The text is divided into four main chapters, while the fifth one summarizes the entire book. The first chapter discusses the interstellar medium, its properties, and its chemistry. The second chapter is dedicated to the characteristics and the dynamical evolution of nebulae. The third chapter discusses the concept of triggered star formation, while the fourth one serves as short introduction to computational astrophysics presenting examples of expanding nebulae. The reader should be aware that computational astrophysics is a subject that evolves rapidly, and consequently the literature is updated extremely frequently (approximately on a weekly basis).

At this point I would like to thank various colleagues with who I had interesting discussions for many aspects of that book. I would like to thank Thomas Haworth, Serena Viti, Padelis Papadopoulos, Mike Barlow, Jeremy Yates, Patrick Owen, Jonathan Mackey, and Robin Williams. Special thanks go to my Ph.D. supervisor Anthony Whitworth who was the first to introduce me to computational astrophysics and also to my colleagues Richard Wünsch, Jan Palouš, Stefanie Walch, David Hubber, Dimitris Stamatellos, and Jim Dale. Thanks also go to various other people:

Dimitra Strongylou, Panagiotis Kotsios, Elena Bacharopoulou, Areti Malkogianni, Anti Nalmpanti, Christophoros Tsantoulas, Antonios Makrymallis, Vasileia Filidou, Angelos Tsiaras, Dimitris and Maria Miroti, Kostas Tziotziou, Dionisis Chandolias, Lazaros Voulgaris, Maria Cvorkov, and Giannis Orkopoulos.

My deepest thanks go to my parents George and Xanthi, my siblings Anna and Elias, my grandparents Elias and Elli, and to Panos for their love and support throughout my entire career. Without them this book would not have been possible.

London, UK   Thomas G. Bisbas
September 2015

Contents

Acronyms

AGN   Active galactic nuclei
BRC   Bright rimmed cloud
C&C   Collect and collapse
CNM   Cold neutral medium
EGG   Evaporated gaseous globule
FUV   Far ultraviolet
GMC   Giant molecular cloud
ISM   Interstellar medium
ISRF   Interstellar radiation field
PAH   Polycyclic aromatic hydrocarbon
PDR   Photodissociation region
RDI   Radiation-driven implosion
SNR   Supernova remnant
WNM   Warm neutral medium
YSO   Young stellar object

Chapter 1
The Interstellar Medium

Abstract This chapter gives a brief introduction on the interstellar medium (ISM). It starts with an overview (Sect. 1.1) of the ISM properties including some worth-mentioning historical events that took place during its discovery and exploration. We then highlight the importance of the dust in ISM (Sect. 1.2) and we give the very basic factors consisting the interstellar radiation field (Sect. 1.3). The importance of photodissociation regions is also addressed (Sect. 1.4) which is one of the key points in understanding the chemistry of the ISM and the star formation process therein. Although there are many factors controlling the thermal balance of the ISM chemistry, we focus on one of them which is cosmic rays (Sect. 1.5). Methods detecting molecular hydrogen are also addressed (Sect. 1.6) and we close this chapter by describing the life cycle of the stars and the ISM (Sect. 1.7).

1.1 Overview

At an altitude of 4200 m above sea level, the observatories in Mauna Kea in the Big Island of Hawaii are one of the most spectacular astronomical sites on Earth. At this high altitude, the low levels of humidity and dust along with the crystal clear atmosphere particularly during moonless nights, allows us to observe our true origin: the Universe. Standing naked-eye in front of some thousands of stars, one can identify constellations, open star clusters, some of the planets of our solar system as well as more 'spontaneous' effects of the sky such as meteors. This site has fascinated mankind from the very beginning, with Babylonians and Ancient Greeks to create the most astonishing sky mythology. In one of them, Zeus and the mortal woman Alcmene had a son, Heracles. Zeus let Heracles to suckle Hera's milk, who was his divine wife. But Hera was asleep and when she woke up she pushed away Heracles, being an unknown infant to her. As a result of the push, the milk was spread in the sky along a path, which is the so-called *Milky Way*.

Our Galaxy,[1] the Milky Way, is visible to the naked eye as a bright stream of starlight which when observed with a small telescope or large binoculars, countless stars along this path are revealed (see Fig. 1.1). However, these stars only constitute a

---

[1]The capitalized word Galaxy indicates our galaxy.

© Springer International Publishing Switzerland 2016
T.G. Bisbas, *The Interstellar Medium, Expanding Nebulae and Triggered Star Formation*, SpringerBriefs in Astronomy, DOI 10.1007/978-3-319-26142-3_1

small fraction—about 25 %—of the total mass of our Galaxy and to be more precise, the total *baryonic* mass of our Galaxy. The rest 75 % is the so-called *Interstellar Medium* (ISM) which is invisible to the naked eye. The ISM is the matter that pervades the space between stars in our Galaxy and in general in all galaxies of the Universe. In Milky Way, the ISM is found to be 99 % in gas phase and 1 % in dust, a ratio that astronomers often call *gas-to-dust ratio*.

The ISM is a very rarefied medium for the Earth standards. Unfortunately in the science of Astronomy it is very hard to give examples from the everyday life that can potentially match a quantity corresponding to a Universal scale (such as distance, density, temperature or other quantities). However, we will try to describe how rarefied the ISM is in a simple way. Imagine that you hold a 500 ml glass which does not contain any liquid. For the Earth standards, we call this an 'empty' glass. For the standards of Astrophysics that we examine here however, it is full of atmospheric air. The density of the atmospheric air at the sea level is about $\sim 10^{20}$ particles cm^{-3}, corresponding to $\sim 1.2\,kg\,m^{-3}$. This means that within 500 ml there will be $\sim 5 \times 10^{22}$ particles cm^{-3}. If we were to fill up this glass with ISM, that number would dramatically decrease

(continued)

Fig. 1.1 The Milky Way. In clear nights and away from light-polluted areas, our Galaxy is visible to the naked eye as a stream of faint light. This photo was taken in Filippaious of Grevena, Greece. Photo by Kallias Ioannidis

down to just 500 particles. This density of $n_{\mathrm{H}} \sim 1\,\mathrm{cm}^{-3}$, where n_{H} is the H-nucleus number density, is the average density of the ISM found in our Galaxy.

Due to the size of our Galaxy being $\sim 30\,\mathrm{kpc}$ across and $\sim 0.6\,\mathrm{kpc}$ wide (Rix and Bovy 2013), this very rarefied ISM accumulates in turn a huge amount of mass which is approximately $\sim 5 \times 10^9\,\mathrm{M_\odot}$ (where $\mathrm{M_\odot} = 1.989 \times 10^{33}\,\mathrm{g}$ is one solar mass) corresponding to $\sim 5\,\%$ of the total Galactic mass. The ISM cannot be seen naked eye even in the most clear nights anywhere on our planet. It therefore remained missed for many centuries until the beginning of 1900s. In 1904, Johannes Hartmann was the first to identify interstellar absorption of the K line of calcium emitted from the Delta Orionis star (known as 'Mintaka located in Orion's belt). This absorption could only be explained due to the interaction of that light with an interstellar cloud somewhere along its line of sight. Hartmann's important discovery of 1904 gave birth to the study of the ISM which was reached by a manmade object, Voyager 1, for the first time in August 2012.

Several studies of the chemical composition of the ISM in the Milky Way reveal that it consists of $\sim 70\,\%$ of hydrogen (H), $\sim 28\,\%$ of helium (He) while the rest $\sim 2\,\%$ of heavier[2] elements. It is very interesting to note at this point that the nucleosynthesis occurred during Big Bang (known as 'Big Bang nucleosynthesis') showed that shortly after the creation (or "appearance") of the Universe, the chemical composition of Cosmos consisted of 75 % of hydrogen, 25 % of helium, and *no* heavy elements. Those heavy elements formed during the evolution of the Universe and in particular inside the cores of massive stars. The first stars ever formed after Big Bang are known as Population III stars (see Glover 2013 for a review) and they are thought to be extremely metal-poor thus formed only by hydrogen and helium.

Inside the core of a star however, the temperature is very high allowing the protons to reach high speeds and approach each other by distances less than $10^{-13}\,\mathrm{cm}$. For distances less than this barrier point, the attractive nuclear forces are much stronger than the repulsive electrical forces, which ignite nucleosynthesis processes (known as 'stellar nucleosynthesis') forming heavier elements. It has been found that inside the core of a star, elements up to (and no heavier than) iron-56 ($^{56}\mathrm{Fe}$) are able to form. Once the lifetime of a star comes to an end—and if massive enough i.e. $M > 8\,\mathrm{M_\odot}$—it dies catastrophically undergoing a supernova explosion. It is in this explosion that nucleosynthesis (known as 'supernova nucleosynthesis') allows the formation of heavier metals than $^{56}\mathrm{Fe}$ which then enrich the ISM. Hence, the next generation of stars were more rich in metals. For example, Population II stars are more rich in metals than Population III stars as they contain remnants of the latter. Since the Big Bang, this lifecycle continued until we reach the present

---

[2]In the astronomical nomenclature, a heavy element corresponds to anything but hydrogen and helium.

day chemical composition of the ISM observed in our Galaxy. The proportion of metals in comparison with hydrogen and helium is called *metallicity* and it is most commonly marked as Z.

In Astrophysics we define as *solar metallicity* ($Z_\odot$) the proportion of metals in comparison with hydrogen and helium for the environment of our Sun. Let us consider this value to be unity, therefore $Z = 1\,Z_\odot$ is the solar metallicity (see Asplund et al. 2009 for a review). Studies have shown that the metallicity of the ISM in Milky Way can *on average* be well approximated with the solar value. From the discussion above it turns out that Z is expected to be $Z < 1\,Z_\odot$ in most extragalactic objects, that is objects outside Milky Way, particularly those in the Early Universe. It also turns out that Z increases in time since more and more supernova explosions are occurring, the remnants of which enrich the interstellar medium with higher abundance of metals than it previously had.

It should be interesting to consider an example showing the abundances of species using Earth standards. To do this, let us assume an adult male whose weight is 80 kg. Biologists have found that on average 65 % of the total body weight is constituted by water (H_2O). Therefore, the total mass of water of that adult is 52 kg. Taking into account that one oxygen atom is 16 times heavier than one hydrogen atom, the total fraction of hydrogen mass of a water molecule is 1/8. Hence 6.5 kg of the adult's total water mass is pure hydrogen, which is approximately 8 % of his total body mass. This 8 % comes directly from material formed during Big Bang while almost all the rest 92 % is a product of stellar and supernova nucleosynthesis. In other words, we can state that a human body is mainly consisted of remnants of a dead star.

The ISM is known to appear in different phases according to its density and gas temperature. It can appear in molecular, atomic, and ionized form. Early work by Field et al. (1969) showed that the atomic gas can appear in two phases in thermal equilibrium: cold ($T_{gas} \sim 50\text{--}100$ K) and dense ($n_H \sim 20\text{--}50\,\text{cm}^{-3}$) known as 'Cold Neutral Medium', or warm ($T_{gas} \sim 6 \times 10^3\text{--}10^4$ K) and diffuse ($n_H \sim 0.2\text{--}0.5\,\text{cm}^{-3}$) known as 'Warm Neutral Medium' (WNM). In case the gas has a temperature of $100 < T_{gas} < 10^4$ K it is unstable and depending on the density will either turn to CNM or WNM. About 8 years later, McKee and Ostriker (1977) extended the two-phase to three-phase model by considering the contribution of supernova explosions, the energy of which is enough to maintain a hot ($T_{gas} \sim 10^6$ K) and rarefied ($n_H \sim 10^{-2}\,\text{cm}^{-3}$) medium known as 'Hot Ionized Medium' (HIM).

The gas in the Universe has been slowly accreted to form galaxies. This material found between galaxies is called 'intergalactic medium'. We believe that even at present day our Galaxy grows in mass by accreting material via a continuous flow of intergalactic medium to the Halo of Milky Way. This new gas is very hot but once it enters our Galaxy it cools down forming the ISM. Overall, the ISM constitutes approximately 3 % of the total composition of the Universe and stars constitute its

1 %. The rest 96 % of the composition of the Universe is constituted by dark matter (27 %) and by dark energy (69 %) the descriptions of which are beyond the scope of this book.

1.2 The Dusty ISM

The ISM of the Milky Way is observed to contain large dark patches, particularly when seen in visible wavelengths. These patches, known also as 'dark nebulae', are opaque enough to block out the background visible light. Observations have shown that these objects have densities of $n_H \sim 10^2$–$10^4 \, cm^{-3}$ and gas temperatures of $T_{gas} \sim 10$–100 K. Isolated small dark nebulae are also known as "Bok globules". Under such conditions, the formation of H_2 is possible, and as we will see in the following chapters, these objects are also potential sites for star formation to occur. Figure 1.2 shows such a dark nebula. If we were to observe a dark nebula at infrared wavelengths—which are longer than the visible—then the background stars could be seen, making it transparent. The infrared light cannot penetrate Earth's atmosphere hence we need space telescopes to perform such observations. One such infrared space telescope was *Herschel* which provided to community an incredible gallery and observational data of various objects. We may therefore state that a dark nebula consists of material effective enough in absorbing the visible light, thus having the size of the visible wavelength, i.e. $\lesssim 1 \, \mu m$. This material is *interstellar dust particles* and it constitutes on average ~ 1 % of the ISM.

The origin of interstellar dust is unclear but it is believed that it is a product of supernova explosions. Observations of the SN 1987A supernova, point to the fact that dust particles have been formed about 530 days after the explosion (Danziger et al. 1989). However, the highly supersonic blast of that explosion is able to completely destroy grains during fast expansion hence a supernova can immediately become a destructive mechanism of interstellar dust. It is alternatively thought that grains could also have been formed in the atmosphere of cool supergiants, the stellar wind of which can then release those grains in the environment of the ISM. Since grains are present in such environments, we may assume that a dust grain contains in fact very heavy elements. Indeed, the *STARDUST* mission which was launched in 2006, brought back to Earth about eight cosmic particles revealing the presence of Al, Fe, Mg, Si and others. It is interesting to note that one of the most astonishing discoveries of that mission was that the amino acid glycine—which is a fundamental building block of life—was detected in the cometary tail of a comet (Wild-2) meaning that the presence of life in the Universe may not be as rare as we think.

A dust grain has on average a diameter of about 500 nm, while the distribution of sizes, r_{dust}, follows a $r_{dust}^{-3.5}$ power law. This ~ 500 nm size is appropriate for Rayleigh scattering to occur (giving the blue colour in the sky) and in which the blue light (shorter wavelengths) is scattered more than the red light (larger wavelengths). This effect is responsible for the formation of reflection nebulae which have been observed in the ISM. Reflection nebulae are objects which indeed have all a bluish colour resulting from the Rayleigh scattering. A well known reflection nebula is Pleiades (see Fig. 1.3). Furthermore, once the electromagnetic radiation has been reflected, it is immediately prone to polarization. Polarized starlight has also been observed (i.e. White 1979). In addition, an interesting consequence of the Rayleigh scattering is the interstellar reddening. Here, in case an object has been interposed by relatively high amounts of dust, it appears to be more red (hence the word 'reddening') as all other wavelengths—and mainly the blue one—have been scattered allowing only wavelengths $\gtrsim 500$ nm to propagate unaffected.

Various observations show that the elements in the gas phase, particularly the metals, are by factors of $\sim 10^2$–10^3 under-abundant relative to hydrogen comparing to solar abundances. This effect is due to depletion of these elements from the gas phase onto the interstellar dust grains which coexist with gas. Although dust accounts for $\sim 1\%$ of the total ISM yet it plays a significant role in determining the thermal balance of the latter. For example, a dust grain contributes in the formation of H_2 in a twofold way (Cazaux and Tielens 2002, 2004). First, it can act as a catalyst; if a hydrogen atom 'lands' on the surface of a dust grain, it can stay there for relatively long time as well as it can 'walk' along this surface. In case another hydrogen atom lands, it can recombine with the other atom forming H_2. Upon this recombination, energy is released which is able to shoot out the newly

Fig. 1.2 The horsehead nebula (IC434) located in the constellation of Orion. This dark nebula (enlarged embedded image) blocks out the background visible light particles which shield radiation. Filters used: $H\alpha$, $H\beta$, blue, green. Photo taken at the Machon Observatory Panorama Thessaloniki by Kallias Ioannidis

formed H_2 molecule enriching the medium. It is interesting to note here that due to this mode, knowledge of the surface of a dust grain is important. A dust grain is far from spherical and instead it is elongated following some fractal morphology thus changing the surface-to-volume ratio. Second, a dust grain blocks out the UV radiation as we have explained above, making high optical depths UV-free (i.e. at the inner parts of a cloud), creating an environment appropriate for H_2 formation.

Another important component of interstellar dust is the polycyclic aromatic hydrocarbon (PAH) molecule. A PAH is an organic two-dimensional molecule consisting of carbon and hydrogen only. PAHs play an important role in the ISM heating (Wolfire et al. 2003, 2008). When a UV photon carrying energy $h\nu \geq 6\,eV$ interacts with a PAH, an electron is liberated as a result of the photoelectric effect. The surplus of the energy of the photon that has interacted with the PAH is absorbed by the newly freed electron ('photoelectron') in the form of kinetic energy. While

Fig. 1.3 Pleiades (M45). This reflecting nebula consists of dust particles which are illuminated by light from nearby stars. Due to the size of the dust particles being similar to the wavelength of the blue light, only this component is reflected whereas higher wavelengths simply pass through the region without suffering any scatter effect. Filters used: RGB. Photo taken at Moustheni Kavala by Kallias Ioannidis

PAHs are organic molecules and are observed even in the Early Universe, it is suggested they play one of the key roles in our understanding of how the abiotic life has been formed.

1.3 Interstellar Radiation Field

Consider a small box with volume of $1\,cm^3$ which is placed at a random position inside the Milky Way, but also away from high-energy and high-density objects such as a supernova or a star forming region. Apart from interstellar gas and dust as we discussed earlier, this box will contain also energy. This energy is coming from different energetic components which all together create the so-called *interstellar radiation field* (ISRF). Similarly to the 'gas density', the 'energy density' describes the amount of energy that is included in this volume of $1\,cm^3$ and in the cgs system it has units of $erg\,cm^{-3}$.

The ISRF changes in places inside the Galaxy as well as in any of the extragalactic objects. Based on detailed measurements made in the solar neighbourhood (the so-called 'local IRSF'), Draine (2011) provided energy densities for six different components of the ISRF. According to these values, the component of the radiation emitted by nearby stars (starlight) accounts for the $\sim$53.2 % of the total energy density included in the $1\,cm^3$, dust emission accounts for $\sim$25.3 %, cosmic background radiation for $\sim$21.2 %, while the rest is covered by nebular emission $\sim$0.3 % with trace amounts of energy due to soft X-rays and synchrotron radiation $\ll$0.1 %. Figure 1.4 shows a Mondrian box[3] the partitions of which correspond to the fraction of each component consisting the local ISRF.

The energy density due to the starlight component was examined in detail by Habing (1968) and later by Draine (1978). Habing (1968) estimated an energy density of $5.29 \times 10^{-14}\,erg\,cm^{-3}$ which corresponds to photons carrying an energy between 6–13.6 eV and which are responsible for the formation of Photodissociation Regions (see Sect. 1.4). Astrophysicists examining the ISRF in the ISM usually adopt the G_o unit called "Habing" and which is equal to the normalized value of energy density with $5.29 \times 10^{-14}\,erg\,cm^{-3}$. About a decade later, Draine (1978) found a different starlight energy density being equal to $1.69\,G_o$. Similarly, this latter unit is also widely known as the "Draine" unit.

1.4 Photodissociation Regions

As we have expressed previously, the radiation emitted by stars plays a key role in determining the phase of the ISM as to whether or not it is ionized, atomic, or molecular. The interaction of the strong UV radiation ($h\nu \gtrsim 13.6$ eV) with a

---

[3]The term 'Mondrian box' adapted in this book is due to the similarity of Fig. 1.4 with paintings made by Piet Mondrian (1872–1944).

cloud is able to ionize part of it. As this radiation propagates further into the cloud, it attenuates reaching energies between $6 < h\nu < 13.6\,\mathrm{eV}$. This far-ultraviolet radiation (FUV) is predominantly responsible for the chemical state of the neutral gas, creating regions commonly referred to as 'Photodissociation regions' (PDRs; known also as 'photon-dominated regions'). A PDR is the in-between interface layer of this atomic gas connecting the ionized and molecular chemical states of the ISM. The study of PDRs is important as we gain a much deeper understanding of the effects of the FUV photons in governing the physical and chemical structure and determining the thermal balance of the neutral ISM in galaxies.

The bulk of the ISM is in neutral form. Compromising the WNM, CNM and molecular clouds that are the birthplaces of stars, the study of PDRs can reveal useful and vital information about the conditions within star-forming regions. PDRs are also responsible for the majority of the infrared emission observed in galaxies. Information about the gas and dust within a galaxy is encoded in that emission, which can pave the way towards our deeper understanding of those distant objects. PDRs cannot only be formed from the interaction of the ISM with starlight. FUV photons and strong X-ray emission can also be emitted by the accretion process of material onto supermassive black holes in active galactic nuclei (AGNs). This radiation can in turn interact with the surrounding gas of those objects thus creating extended PDRs, much bigger than those found around massive stars. In addition, supernovae explosions do enrich the surrounding ISM with metals which therefore change (increase) its metallicity and which has an impact on the overall gas and dust temperature profiles. Examining these extreme environments provides useful insight into the dynamics of non-local ISM observed at the centers of merger and starburst galaxies.

One of the most known examples of PDRs is the Orion Bar. The Orion Bar was one of the first objects to study to better understand the FUV photon chemistry. Reproducing the PDR chemistry of that object has been attempted by the theoretical models suggested by Tielens and Hollenbach (1985a,b) who examined the reactions followed in a dense gas as it is exposed to high FUV fluxes. Their study gave rise to a series of other theoretical PDR studies while during the last 20 years or so more detailed models have been introduced along with complex numerical algorithms that are able to solve the chemistry for large networks containing hundreds of species and thousands of chemical reactions. In particular, Röllig et al. (2007) presented a comparison study of 11 different PDR codes. Until then further similar codes have been developed focusing particularly on tackling three-dimensional density distributions (Bisbas et al. 2012; Andree-Labsch et al. 2014) including also hydrodynamics (Glover et al. 2010).

Figure 1.5 shows a diagram of the basic structure of a PDR. The radiation is impinging from left to right. From this sketch we see that a PDR contains chemically-distinct zones. For a better understanding of those, we can divide them in groups which contain hydrogen, carbon, and oxygen. At the outermost part of the PDR i.e. where the FUV radiation is impinging from, hydrogen appears primarily in its atomic phase (H I). This occurs since all photons above the Lyman limit are

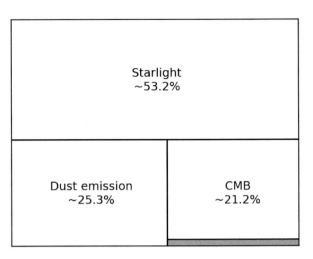

Fig. 1.4 The Mondrian box showing the three dominating components of the local ISRF and the corresponding fraction. Each partition is scaled according to that fraction. The *bottom right shaded region* shows the total contribution of nebular emission with the trace amounts of energy due to soft X rays and synchrotron radiation

absorbed and hence electrons and protons recombine to form a hydrogen atom. In addition, molecular hydrogen is also photodissociated since there is enough surplus of FUV radiation to destroy its molecular state. Carbon is found in its singly ionized phase (C II), as an FUV photon is able to remove one electron from the carbon atom. C II extends at a greater depth than the H I zone due to the self-shielding of H_2 thus its layer is more thick. Similarly, oxygen is found in its atomic phase (O I) as the FUV photons are able to dissociate O_2. On average, the temperature of the gas in this outermost part of a PDR is approximately $\sim$100–200 K, although models and observations show that it can be as high as $\sim 10^3$ K depending on the intensity of the FUV radiation and the density of the gas.

As the FUV is extinguished when one moves deeper in the cloud, the formation of H_2 starts to dominate over photodissociation. The area where the abundance of H_2 is higher than that of H I is known as 'H I-to-H_2 transition layer' and it is an area where most of the studies (at least in the past) have been concentrated. Even deeper in the cloud, the FUV becomes so severely extinguished that is negligible. Carbon makes the transition from a singly ionized state to being locked in the stable molecule carbon monoxide (CO) (van Dishoeck and Black 1988). Between these two states, the carbon is in neutral atomic form and can attain high abundances in the transition layer beyond the C II zone. The gas temperature has now dropped down to $\sim$10 K unless other heating sources found in these large optical depths can maintain it to higher values. These heating sources can be cosmic rays (see Sect. 1.5), turbulent heating which comes from the hydrodynamical motion of the cloud, or shock heating which comes from the propagation of a shock front.

Observationally, PDRs are very important as they are the primary sources of the emission of spectrum lines which are captured by various ground-based (i.e. ALMA, JVLA, JCMT), air-borne (i.e. BLAST, SOFIA), and space-borne (i.e. *Herschel*, *Spitzer*) telescopes. These lines are the lines emitted by C II at 158 μm corresponding to the $^2P_{3/2} \rightarrow ^2 P_{1/2}$ transition, by C I at 609 and 370 μm

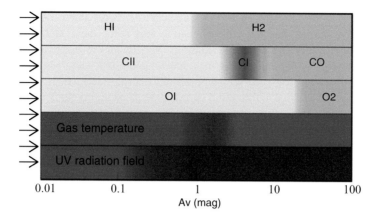

Fig. 1.5 A sketch showing a PDR. The radiation is assumed to impinge from *left* to *right* as the *arrows* indicate. The logarithmic *x*-axis corresponds to optical depth. The *bottom tier* shows the extinction of the UV radiation (*red* is high-UV, *black* is low-UV) and *above* it, the gas temperature is shown (also *red* is high and *blue* is low). The *top three tiers* correspond (from *bottom* to *top*) to oxygen (transition from O I to O_2), to carbon and carbon monoxide (from C II to C I and then to CO), and to hydrogen (from H I to H_2). Note that C I is primarily located at intermediate optical depths and that as we go from low to high optical depths, the atoms recombine to form molecules

corresponding to $^3P_1 \rightarrow\, ^3P_0$ and $^3P_2 \rightarrow\, ^3P_1$ respectively, by O I at 63 and 145 μm corresponding to $^3P_2 \rightarrow\, ^3P_1$ and $^3P_1 \rightarrow\, ^3P_0$ respectively, and by the various CO J-transitions such as $J = 1 - 0, J = 2 - 1$ and higher.

1.5   Cosmic Rays

Cosmic rays were discovered by Viktor Hess during balloon flights in 1912–1913. He described them as "extremely penetrating radiation coming from above the atmosphere". However, the term 'radiation' (or 'ray' in 'cosmic rays') is probably misplaced. In fact, cosmic rays are high energetic *particles* with energies of the order of several hundred MeV whereas their energy density in Milky Way is estimated to be $\sim$1 eV/cm^3. They move at speeds close to the speed of light and because of that they are relativistic particles. Cosmic rays are in principle consisting of $\sim$90 % hydrogen nuclei (protons), of $\sim$9 % helium nuclei, while the remaining fraction corresponds to heavier nuclei than helium. A tiny fraction of cosmic rays consist also of relativistic electrons and positrons (positively charged electrons). However, this much lower flux of electrons and positrons can provide useful information about their origin. Since cosmic rays carry charge, they follow the magnetic field lines of the entire Galaxy and are therefore bound to it. In general, every particle moving at such high energies can be considered as a cosmic ray particle.

The origin of cosmic rays in our Galaxy is not yet well understood. However, it is strongly believed that the bulk of cosmic rays reach these high energies by accelerations in supernova remnants. During a supernova explosion, a tremendous amount of energy is released which may reach up to $\sim 10^{51}$ erg. Thus the ejected material of the dying star gain very high kinetic energies and which create highly supersonic speeds relative to the sound speed of the surrounding (undisturbed) ISM. This effect results in the formation of an expanding strong shock front. It is in this front that the particles of cosmic rays are being accelerated, as during the expansion head on collisions take place between the particles and the shocked material. It is immediately intuitive to consider that the faster the shocked material moves, the more the particles gain energy. This mechanism is known as *diffusive shock acceleration* and is considered to be a case of a more general idea initially proposed by Fermi in 1949. It can explain the majority (reaching $\sim 90\%$) of the observed energy distribution of cosmic rays.

Cosmic rays are able to penetrate and even ionize the gas at a *statistically* constant rate known as the "cosmic ray ionization rate" and which is denoted as ζ_{CR}. Each galaxy has different ζ_{CR}, with the ones found in the Early Universe being more than 500 times the rate observed in our Galaxy. This rate can also dramatically change in environments close to supernova explosions, such as in the Crab Nebula. The average value of ζ_{CR} in Milky Way is $\zeta_{CR} \simeq 1.3 \times 10^{-17}\,\mathrm{s}^{-1}$ (i.e. Williams et al. 1998), although Dalgarno (2006) has proposed a higher value of $\zeta_{CR} \simeq 10^{-16}\,\mathrm{s}^{-1}$.

What is the importance of cosmic rays in astrochemistry? The answer is simple and it is that "without cosmic rays some species wouldn't exist in places where radiation is shielded". The reason for this is that as they penetrate deeply inside the cloud, they react with H_2 which has been formed and which create the reactions

$$H_2 + CR \longrightarrow H_2^+ + e^-,$$
$$H_2^+ + H_2 \longrightarrow H_3^+ + H,$$

and the latter one reacts with the free e^- creating various other products following

$$H_3^+ + e^- \longrightarrow \text{several products},$$

which actually ignite a large chemical network.

1.6 Detection of H$_2$

As we have described above, the bulk of the baryonic non-stellar mass of the Universe consists of molecular hydrogen (H$_2$) with a total fractional abundance of $\sim$70 %. It is this gas that regulates the evolution of galaxies across the epoch times. Molecular hydrogen is responsible for star formation and in general plays a fundamental role in many astrophysical contexts. H$_2$ is found in all cold ($T \sim 10$ K) and dark regions where the UV photons emitted by stars do not penetrate.

However, H$_2$ is not visible in optical wavelengths and does not emit radiation which can be captured by radio telescopes. In particular it lacks a permanent dipole moment and can only change ro-vibrational state through weak quadrupolar transition with high excitation temperatures $\gtrsim$500 K that are sub-thermally populated and so difficult to detect in the cold bulk of the ISM. Furthermore, the pure rotational transitions of H$_2$ lie at mid-infrared wavelengths that are largely inaccessible from the ground, due to telluric absorption and strong background emission. Because of this, astronomers have implemented techniques which are used to observe this gas indirectly. One of the most common techniques is the one using carbon monoxide (CO), also very abundant in the Universe, as a tracer. CO is formed in places where H$_2$ also forms and is visible in radio telescopes. Once an accurate measurement of CO emission is made, astronomers apply as simple factor (called the X_{CO}-factor) that converts this emission to H$_2$ gas mass. This X_{CO}-factor is defined as the column density of the molecular hydrogen along a line of sight divided by the velocity-integrated intensity of the ^{12}CO(1-0) rotational transition line. The relation is

$$X_{CO} = \frac{N(H_2)}{\int T_A(CO)dv} \tag{1.1}$$

where N(H$_2$) is the column density of H$_2$ and T_A(CO) is the antennae temperature of the ^{12}CO $J = 1 - 0$ transition line. For molecular clouds in the Milky Way, $X_{CO} \sim 2 \times 10^{20}$ cm^{-2} K^{-1} km^{-1}s (Strong and Mattox 1996; Dame et al. 2001). This has now become the standard method for inferring molecular gas content and has been found to work reasonably well in the ISM of our Galaxy (a review on the CO-to-H$_2$ conversion factor is discussed by Bolatto et al. 2013). Recently, Clark and Glover (2015) found that the X_{CO} factor may depend on the star formation rate particularly in extragalactic objects.

There are, however, other alternative methods to trace H$_2$ in the Universe using atoms and molecules other than CO. The atomic carbon C I, which has been firstly proposed by Papadopoulos et al. (2004) as an alternative tracer to CO, becomes now one additional widely used method. This study has been further examined by Offner et al. (2014) and Glover et al. (2015) with some first observations in the Vela Molecular Ridge cloud C supporting its confirmation (Lo et al. 2014). Other tracers of H$_2$ include CH (Gredel et al. 1993) and the HF molecule (Sonnentrucker et al. 2010; Neufeld et al. 2010) with the latter to be one of the best options.

A recent theoretical model proposed by Bisbas et al. (2015), suggests that cosmic rays play a major role in tracing H_2 in the Universe. They found that ordinary boosts of cosmic rays efficiently destroy the molecule of CO but not H_2. Such ordinary boosts of cosmic rays are expected (and observed) in various star-forming galaxies where star formation rate (i.e. the rate at which stars are being formed in unit time) is high, or in most of the Early Universe galaxies. The main effect proposed is that a cosmic ray particle forms singly ionized helium (He II) which reacts and destroys almost at 100 % efficiency the CO molecule producing C II and O. While under such circumstances H_2 is able to survive, tracing it using CO is in practice not possible. CO (as well as H_2) however can survive at high densities even during high cosmic-ray ionization rates making the galaxies to appear more clumpy than they truly are. Instead, an alternative method for tracing H_2 in such environments is that of C I as suggested by Papadopoulos et al. (2004).

1.7 Life Cycle of the Stars and the ISM

As it inevitably occurs with every part consisting the Universe and whether this is biological or not, the stars and the ISM have a life cycle. In general, the life cycle is a sequence of repetitive stages that evolve a system in time. For our Galaxy as well as for all other galaxies in our Universe, this is determined by the evolution of the ISM as it forms stars which in turn form ISM again following various different processes.

As we will explore further in this book, under some conditions the ISM may become gravitationally unstable and collapse to form a new star or a small cluster of stars and other objects satisfying hydrostatic equilibrium. As those stars evolve and depending on their mass, they eject the material back to the ISM either with stellar winds or undergoing a catastrophic explosion leading to supernovae. In both cases and particularly in the second one, the ejected material has suffered large pressures while it was inside the star and its chemical stage has thus changed with the metals to be of higher abundance than before. Hence, we may claim that during the life cycle of ISM, its metallicity increases while at the same time most of its material responsible for the formation of the dying star has been returned back to its origin. The rest of that material remains with the extremely dense core of the dead star which may be a white dwarf, neutron star, or a black hole. Therefore, if we suppose that M_{init} mass of the ISM was fully converted to one single star, after the death of the star $x \cdot M_{init}$, where $x < 1$ has been converted to ISM again. Other low mass objects (i.e. $M < 0.08\,M_\odot$) which do not ignite nuclear reactions act to lock up ISM gas since they are quite stable and their lifetime may comparable to the lifetime of the Universe.

In other words, the total mass of the ISM found in one galaxy decreases in time while it becomes more rich in metals. In timescales comparable to the timescale of a galaxy's lifetime (or even larger), the above process results to the decrease of the ISM gas which is inevitably connected with the star formation rate (that is the rate at which star formation occurs), thus on average constantly decreasing also the luminosity of the entire galaxy.

While metallicity increases, one would expect that older stars have less metals than those that are currently forming. As we have referred to previously, astronomers use the term "Population" to determine the different kinds of stars. Two populations of stars are widely used: Population I and Population II. Population I stars are more metal rich than those in Population II. The hypothetical Population III stars which are completely metal poor are thought to exist in the Early Universe and due to their lifetime one cannot observe such a star in our Galaxy (our Galaxy is too 'old' to host such stars). Therefore, one would need high resolution observations in the very distant Universe in order to be able to identify them. Indeed, during the time that this book was written, an ESO[4] team of astronomers using the VLT array were able to discover such a distant galaxy (namely 'CR7') which, to date, is the best candidate for hosting Population III stars (Sobral et al. 2015).

Examining the lifecycle of the ISM helps us to understand how the whole cyclic process of gas-to stars-to gas occurs. However, this is a difficult task and requires the contribution of the entire community as each different group studies in detail different particular aspects of this cycle. This is perhaps a side, yet important, consequence of studying such large scale events: the collaboration between a large group of scientists that can lead to the answer of the most astounding question "what are the procedures followed to create conditions that can ignite life?"

Acknowledgements I would like to thank Dr. Padelis Papadopoulos and Dr. Nick Prantzos for the useful discussions. In particular I would like to acknowledge Dr. Papadopoulos for the discussions on cosmic rays and Dr. Prantzos for his great outreach idea to explain what is the fraction of the human body that comes directly from the Big Bang and what is the fraction that has been produced inside stars.

References

Rix, H.-W., & Bovy, J. 2013, A&A Rev., 21, 61
Glover, S. 2013, Astrophysics and Space Science Library, 396, 103
Asplund, M., Grevesse, N., Sauval, A. J., & Scott, P. 2009, ARA&A, 47, 481
Field, G. B., Goldsmith, D. W., & Habing, H. J. 1969, ApJ, 155, L149
McKee, C. F., & Ostriker, J. P. 1977, ApJ, 218, 148
Habing, H. J. 1968, Bull. Astron. Inst. Netherlands, 19, 421
Danziger, I. J., Gouiffes, C., Bouchet, P., & Lucy, L. B. 1989, IAU Circ., 4746, 1
White, R. L. 1979, ApJ, 229, 954
Cazaux, S., & Tielens, A. G. G. M. 2002, ApJ, 575, L29
Cazaux, S., & Tielens, A. G. G. M. 2004, ApJ, 604, 222
Wolfire, M. G., McKee, C. F., Hollenbach, D., & Tielens, A. G. G. M. 2003, ApJ, 587, 278
Wolfire, M. G., Tielens, A. G. G. M., Hollenbach, D., & Kaufman, M. J. 2008, ApJ, 680, 384

---

[4]European Southern Observatory.

Draine, B. T. 2011, Physics of the Interstellar and Intergalactic Medium by Bruce T. Draine. Princeton University Press, 2011. ISBN: 978-0-691-12214-4,

Draine, B. T. 1978, ApJS, 36, 595

Tielens, A. G. G. M., & Hollenbach, D. 1985a, ApJ, 291, 722

Tielens, A. G. G. M., & Hollenbach, D. 1985b, ApJ, 291, 747

Röllig, M., Abel, N. P., Bell, T., et al. 2007, A&A, 467, 187

Bisbas, T. G., Bell, T. A., Viti, S., Yates, J., & Barlow, M. J. 2012, MNRAS, 427, 2100

Andree-Labsch, S., Ossenkopf, V., Röllig, M. 2014, arXiv:1405.5553

Glover, S. C. O., Federrath, C., Mac Low, M.-M., & Klessen, R. S. 2010, MNRAS, 404, 2

van Dishoeck, E. F., & Black, J. H. 1988, ApJ, 334, 771

Williams, J. P., Bergin, E. A., Caselli, P., Myers, P. C., & Plume, R. 1998, ApJ, 503, 689

Dalgarno, A. 2006, Proceedings of the National Academy of Science, 103, 12269

Strong, A. W., & Mattox, J. R. 1996, A&A, 308, L21

Dame, T. M., Hartmann, D., & Thaddeus, P. 2001, ApJ, 547, 792

Bolatto, A. D., Wolfire, M., & Leroy, A. K. 2013, ARA&A, 51, 207

Clark, P. C., & Glover, S. C. O. 2015, arXiv:1506.06503

Papadopoulos, P. P., Thi, W.-F., & Viti, S. 2004, MNRAS, 351, 147

Offner, S. S. R., Bisbas, T. G., Bell, T. A., & Viti, S. 2014, MNRAS, 440, L81

Glover, S. C. O., Clark, P. C., Micic, M., & Molina, F. 2015, MNRAS, 448, 1607

Lo, N., Cunningham, M. R., Jones, P. A., et al. 2014, ApJ, 797, L17

Gredel, R., van Dishoeck, E. F., & Black, J. H. 1993, A&A, 269, 477

Sonnentrucker, P., Neufeld, D. A., Phillips, T. G., et al. 2010, A&A, 521, L12

Neufeld, D. A., Sonnentrucker, P., Phillips, T. G., et al. 2010, A&A, 518, L108

Bisbas, T. G., Papadopoulos, P. P., & Viti, S. 2015, ApJ, 803, 37

Sobral, D., Matthee, J., Darvish, B., et al. 2015, arXiv:1504.01734

Chapter 2
Nebulae

Abstract This chapter discusses the basic concepts and properties of ionizing nebulae. In Sect. 2.1 we describe hydrogen ionization and recombination occurring in an ionizing region highlighting the on-the-spot approximation which is frequently used by researchers. In Sect. 2.2 we give a brief overview of the categories of nebulae including an example in each case. In Sect. 2.3 we analyze the physics behind photoionization equilibrium and in Sect. 2.4 we provide the basic heating and cooling mechanisms occurring in an H II region. In Sects. 2.5 and 2.6 we discuss about the dynamical expansion of an H II region and how this leads to a strong shock front formation.

2.1 Ionization and Recombination of Hydrogen

Let us suppose there is a cosmic gas which consists of hydrogen. As we saw in the previous chapter, the ISM has various heating sources which increase its temperature. At a fixed density and as the temperature increases, molecular hydrogen dissociates producing atomic hydrogen (H I) according to the reaction

$$H_2 \longrightarrow 2HI - 4.5\,eV \tag{2.1}$$

where 4.5 eV is the energy needed in order to dissociate this hydrogen molecule. In case the gas is close to a source emitting ionizing radiation, the photons consisting this radiation interact with H I and they ionize it according to

$$HI \longrightarrow HII + e^- - 13.6\,eV \tag{2.2}$$

which creates a free electron and a free proton (ion) while absorbing energy equal to 13.6 eV.

In case a free electron and a free proton are close to each other, they may recombine to form a new H I producing either a single (but highly energetic) photon, or multiple (but lower energetic) photons. This is because during this recombination process the electron may recombine (a) directly to the ground state or (b) to any higher state but then cascading down until it reaches the ground state. In the first case, all 13.6 eV that the free electron has in the form of kinetic energy will be

© Springer International Publishing Switzerland 2016
T.G. Bisbas, *The Interstellar Medium, Expanding Nebulae and Triggered Star Formation*, SpringerBriefs in Astronomy, DOI 10.1007/978-3-319-26142-3_2

converted to a photon having the same amount of energy and which corresponds to the UV radiation. In the second case, various other photons may be emitted, some of them in the visible spectrum.

The *recombination coefficient*, α, describes the volumetric amount of recombinations that can be achieved in time and it has units of volume per unit time (i.e. in cgs it is cm^3 s^{-1}). When multiplying this coefficient with the number of electrons and the number of protons (i.e. $n_e n_p \alpha$ which is simply $n_p^2 \alpha$ for H), we obtain the *recombination rate* per unit volume per unit time. The total recombination coefficient (frequently referred to as "case-A") for a hydrogen atom is given by

$$\alpha_A = \sum_{N=1}^{\infty} \alpha_N(\text{HI}, T), \tag{2.3}$$

where N is the principal quantum number. Note that the recombination coefficient depends on the temperature (the higher the temperature, the smaller the number as it is more "difficult" for a recombination event to occur). This equation takes into account *all* possible recombination states i.e. from $\infty \rightarrow 1$, or $\infty \rightarrow 2 \rightarrow 1$ etc.). Since recombinations *directly* to the ground state can produce a photon capable to ionize a nearby hydrogen atom *diffusively*, it is convenient to neglect this recombination and thus to reduce the complexity of the system of equations when examining the thermodynamics and evolution of nebulae as we will see later on in this chapter. When neglecting that specific recombination (frequently referred to as "case-B"), the recombination coefficient takes the form

$$\alpha_B = \alpha_A - \alpha_1(\text{HI}, T) = \sum_{N=2}^{\infty} \alpha_N(\text{HI}, T). \tag{2.4}$$

The above approximation is well known as the *on-the-spot* approximation (OTS; Osterbrock 1974) and is very commonly used. By invoking the OTS approximation, we simply neglect the contribution of the diffuse component of the radiation. The temperature-dependent value for the case-B recombination coefficient frequently used is $\alpha_B \simeq 2.7 \times 10^{-10} \left(\frac{T}{[\text{K}]}\right)^{-3/4}$ cm^3 s^{-1}. For the ionizing hydrogen at a temperature of $T_i = 10^4$ K, we obtain $\alpha_B = 2.7 \times 10^{-13}$ cm^3 s^{-1}.

> The on-the-spot approximation tells us that *any* region that is shaded will *never* be ionized.

To understand further the effects that may be caused by diffusive radiation, Fig. 2.1 plots a schematic diagram showing such a simplified example. Here, an ionizing source (black dot at the bottom) emits ionizing radiation which interacts with the hydrogen gas around it creating free protons and free electrons (white region). As we will discuss further below (see Sect. 2.6), this ionized region expands

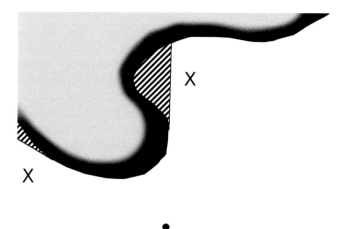

Fig. 2.1 A region interacting with an ionizing source (*black dot*). The *white area* in this diagram corresponds to the region where all hydrogen atoms are ionized. The *black region* corresponds to the shock front occurring due to the expansion of the ionized medium (see Sect. 2.6) and where the material is quite dense and absorbs all ionizing photons leaving all gas behind it undisturbed (*gray region*). The *stripped areas* correspond to the region where the diffuse component of radiation (occurred nearby such as at both places marked with "X") is the only way to ionize it. By invoking the OTS approximation, these *stripped regions* will *never* be ionized

and creates a shock front (black region). The ionizing photons cannot penetrate further and therefore behind the shock front the medium is undisturbed (gray region). The stripped region indicates ionization due to the diffuse component of radiation. Suppose that a recombination event occurs at places marked with "X". If the recombination of the free electron occurs *directly* to the ground state, then an ionizing photon will be produced—as we described earlier—which will be emitted towards a random direction. If this direction is towards the stripped region and interacts with a hydrogen atom therein, it will ionize it *diffusively* i.e. without the latter interacting directly with the ionizing source. By invoking the OTS approximation, we neglect these contributions, therefore under this assumption the stripped region will no longer be able to be ionized. It is therefore important to note also that if a liberated electron does not recombine directly to the ground state, it cannot create a photon capable to ionize a hydrogen atom and is therefore lost.

2.2 Categories of Nebulae

Nebulae[1] are regions which consist of ionized gas, particularly of hydrogen. The source of ionization is usually a single or multiple massive stars. Massive stars, as we have discussed in Chap. 1 emit copious amounts of ultraviolet radiation with

---

[1]Meaning "cloud" in Latin.

photons carrying more energy than the ionization potential of hydrogen (13.6 eV). Suppose that an ionizing photon hits a hydrogen atom. Upon this interaction, the photon is absorbed by the atom which in turn destroys its chemical state by forming a free electron and a free proton. The electron produced by the absorption of these UV photons are called *photoelectrons*. Any surplus in energy exceeding the ionization potential of 13.6 eV will be transferred to the photoelectron in the form of kinetic energy thus increasing the temperature of the ionized region. A characteristic feature of all nebulae is that the temperature in the ionized gas tends to be much higher than the temperature in the surrounding neutral gas. These two gas states are separated by a thin slice known as the *ionization front*. In this slice, the temperature decreases abruptly and so as the chemistry which from the phase of ionization switches to the atomic phase and deeper to the cloud to molecular phase. The difference between the temperatures of the hot ionized and cold neutral gas results in the pressure-driven expansion of nebulae.

Osterbrock (1974) divided ionizing nebulae in three main categories according to their origin and physical characteristics. These are: (1) diffuse nebulae (emission nebulae; Sect. 2.2.1), (2) planetary nebulae (Sect. 2.2.2), and (3) supernova remnants (Sect. 2.2.3). Additionally, reflection and dark nebulae can be also included in the first category (diffuse), however these do not occur due to ionization of the medium as we will see below (Sect. 2.2.4).

2.2.1 Diffuse (Emission) Nebulae

Diffuse nebulae are extended regions (of the order of pc scale) of gas and dust where there are usually embedded stars. If these stars emit UV radiation, they ionize their surroundings creating the so-called "H II regions". These H II regions are also known as *emission nebulae*. They are extended objects spanning a distance of the order of $\sim$10 pc across. They are quite rarefied with typical number densities of the order of 10–100 cm^{-3}. In regions where the youngest massive stars are located, there might be present smaller and more compact H II regions. These *ultracompact* H II regions have been identified around three decades ago (Wood and Churchwell 1989) and they may have a size of the order of a light year or less as well as their densities may even exceed 10^{4} cm^{-3}.

An emission nebula is a possible site of star formation. A famous and well studied emission nebula is the Orion Nebula (M42 or NGC1976). At a distance of $\sim$414 pc (Menten et al. 2007), the Orion Nebula is a high-mass star forming region. Here, newly ionized gas found in the borders of the H II region flows into the main area of the nebula. The Orion Nebula Cluster is the closest massive star formation region to the Earth and is thus best studied. It has been claimed that most of the massive stars have already passed through the main sequence. Chandra X-ray observations (Getman et al. 2005; Preibisch et al. 2005) found $\sim$1000 sources emitting X-rays, which have masses spanning from $\sim$0.1–50 M$_{\odot}$, indicating that this area is one of the richest in X-rays ever observed. Several disks emitting small jets (microjets) and outflows have been identified in the Orion Nebula (Bally et al. 2000). Some of those disks interact from radiation and winds emitted by the nearby massive stars taking a shape of a small cocoon surrounded by a bow shock. Some of those objects create brown dwarfs if they become gravitationally unstable. Brown dwarfs have been detected in the Orion Nebula (Lucas and Roche 2000) and some of them have masses below $0.013 \, M_\odot$ which is the threshold for deuterium-burning.[2]

2.2.2 Planetary Nebulae

Planetary nebulae are formed during the latter evolutionary stages of a star with mass $1 < M < 8 \, M_\odot$. The dying star expands and its outer parts are ejected into the interstellar medium. The remnant core (a white dwarf) emits highly energetic radiation and ionizes this material. An expanding bubble is the result of this process. Inside the ionized region of the bubble, a plethora of cometary knots exists, which are dense clumps surviving the strong ultraviolet radiation. The life time of such a nebula is of the order of a few tens of thousand of years. One of the ongoing topics in the study of planetary nebulae is the origin of the wide range of their shapes. It has been also suggested that a star on the asymptotic giant branch is able to create a magnetic field in a dynamo located at the core-envelope interface. This magnetic field is strong enough to create bipolar or elliptical shapes, as they are widely observed. Since our Sun has mass in this range of masses, its fate is to turn up into a planetary nebula before expanding its size up to Earth's orbit!

The Helix nebula (NGC7293) is a well known and well studied planetary nebula. At its center there is a white dwarf, the remaining core of the initial star, which ionizes a spheroidal region around it. The progenitor star is estimated to had a mass of $\sim 6 \, M_\odot$. Before the planetary nebula formed, it has been suggested it underwent three phases of dredge-up and hot bottom burning (Henry et al. 1999). There are two different rings (inner and outer) with the outer to be less ionized and of less temperature than the inner one. This region has a radius of 2.5–3 pc and contains thousands of photo-evaporating cometary knots (López-Martín et al. 2001). These knots consist of a dense head with an extended—relative to the size of the head— cometary tail. It has been suggested that their origin is due to instabilities in the expanding shock front, which are amplified by the ultraviolet radiation (O'dell and Handron 1996; Meaburn et al. 1998; Burkert and O'Dell 1998). Recent observations have showed also that from the emission lines of the knots examined show strong evidence for the presence of H_2. Matsuura et al. (2009) presented high resolution images of the cometary knots showing their irregular structure and pointed that they have a "tadpole" shape with an elliptical head and bright crescent inside followed by a long tail opposite to the central star.

---

[2]Deuterium burning occurs in stellar systems that have mass $\sim 0.013 \, M_\odot$ and in which deuterium nuclei and free protons react and combine helium-3 nuclei.

2.2.3 Supernova Remnants

Supernovae remnants (SNR) are the remains of the explosion of a massive star. When a massive star with $M > 8\,M_\odot$ approaches its final stages of life, it ejects huge amounts of its initial material into the surrounding medium. This material is expanding at high speeds sweeping up the undisturbed interstellar gas and creating a strong shock front. The remaining core can either result in a fast rotating neutron star (pulsar) or even a black hole. Supernovae are extremely important for understanding the formation and evolution of our Galaxy. They heat up the interstellar medium, distribute heavy elements throughout the Galaxy, and accelerate cosmic rays. They have been also considered to play the dominant role in driving turbulence in the ISM which has significant consequences in the star formation rate of the galaxy (see Walch et al. 2014 and references therein for three-dimensional high resolution simulations showing this).

Supernovae are divided into two main categories: type I and type II. Type I supernovae are characterized by sharp maxima in their light curves which dies away gradually. This class can be subdivided in type Iα which exhibit a strong absorption line due to ionized silicon, and type Iβ which lacks of this line. On the other hand, type II supernovae are distinguished from type I by the presence of hydrogen lines in their spectra. The mass of this progenitor should be at least $\sim$8 $M_\odot$. A recent study by Mackey et al. (2014) showed that the light curve of supernovae may also result from the interaction of a static, pressure-confined shell (resulting from the movement of the massive star into the ISM in association with its emission of winds) with the strong ionizing radiation emitted after the explosion.

A famous (and rather historical) SNR is the Crab Nebula (M1, NGC 1952; see Fig. 2.2). This nebula is the remnant of a supernova explosion observed naked eye by Chinese and Arab astronomers in 1054 AD. It is also the first object ever recorded as the remnants of a catastrophic death of a star. The mass of the progenitor star is believed to have been about $M \sim 9\text{--}11\,M_\odot$ (Davidson and Fesen 1985; MacAlpine and Satterfield 2008). At the center of the Nebula there is a fast rotating pulsar, the remnant of the progenitor star. Using Hubble Space Telescope and Chandra observations, Hester et al. (2002) studied the inner dynamics of the nebula and Lyubarsky (2002) reported that the shock shape in this inner region is highly non-spherical because the energy flux in the pulsar wind decreases towards the axis. On the other hand, the overall structure of this nebula appears to be filamentary due to the Rayleigh-Taylor instability (Hester et al. 1996). Recently, Barlow et al. (2013) have reported the discovery of a noble gas molecular ion (namely the ^{36}ArH$^+$) in the Crab Nebula. It is interesting to note that this was the first ever detection of a noble gas molecule in space and in the particular case of the Crab Nebula it is believed to have formed from the explosive nucleosynthesis in the parent massive star during its core-collapse phase. In addition, Owen and Barlow (2015) showed that the dust observed in the Crab Nebula is closely correlated with the position of the knots and filaments indicating that it is formed inside the supernova remnants and not in the ejecta.

Fig. 2.2 The Crab Nebula (M1, NGC1952). These are the remnants of a catastrophic death of a massive star (supernova explosion) observed naked eye by Chinese and Arab astronomers in 1054 AD. Filters used: Hα, O III, S II. Photo taken at the Machon Observatory Panorama Thessaloniki by Kallias Ioannidis

Both planetary nebulae and SNRs appear to be morphologically different from emission nebulae. That is primarily because of the way they are formed and which is the final stages of a dying star. Both of them, however, appear to be (at least in the early stages of their evolution) bubble shaped[3] and are in general quite small (of the order of a light year across) and while their mass is of the order of $\sim 2\,M_\odot$ or less (depending also on the mass of the parent star). Finally, the temperature of the ionized medium in the surroundings (vicinity) of the dead star may get even twice as that observed in an emission nebula.

2.2.4   Reflection and Dark Nebulae

Reflection and dark nebulae have one common characteristic: their presence indicate the presence of dust. As we have discussed in Sect. 1.2, the ISM consists also of dust particles which have size similar to the wavelength of the blue light. Therefore, if a nearby bright star shines them, they illuminate reflecting only the blue light (since all other longer wavelengths will not be scattered). On the other hand, if a region containing high amounts of dust is in-between a bright region and the observer (i.e. the Earth), then they are able to fully absorb the visible light and they appear as dark

---

[3]Not to be confused with the bubble shaped emission nebulae (H II regions).

areas in the sky. If we place a dark nebula further and further away, stars will start to overlap in the foreground and along the line of sight, so there will be a point which the nebula will not appear as dark anymore. Therefore, in general a dark nebula is a "local" object in this regard. Figure 2.3 shows the ρ-Ophiuchus region which is an example showing simultaneously both a reflection and a dark nebula.

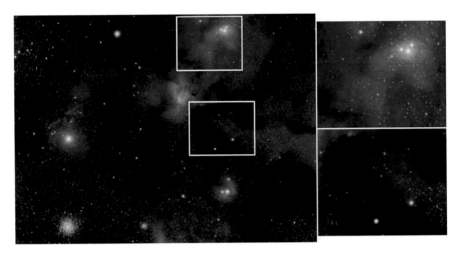

Fig. 2.3 This spectacular image shows the ρ-Ophiuchus region in which one can observe a *bluish* reflection and nearby a dark nebula simultaneously, indicating the presence of high amounts of dust. Filters used: RGB. Photo taken at Thassos island by Kallias Ioannidis

At a distance of $\sim$150 pc, ρ-Ophiuchus is a star forming region in which many young stellar object have been identified (i.e. Nutter et al. 2006). The system is known to contain a brown dwarf with a temperature of $\sim$3000 K which was formed by the gravitational collapse of the surrounding gas. Three-dimensional hydrodynamical simulations by Stamatellos et al. (2007) including radiative transfer techniques showed that the gas temperature of the dark cloud (lower framed image of Fig. 2.3) is colder than $T \sim$ 10–11 K.

2.3 Photoionization Equilibrium

In this book we will focus on the evolution and the characteristics of emission nebulae, since in planetary nebulae and supernova remnants the corresponding physics is different from what we intent to present here and also their chemistry is much more complicated. Consider an arbitrary density distribution $\rho(\mathbf{r})$ consisting of neutral hydrogen. The corresponding number density of hydrogen nuclei in all forms (atomic and molecular) is $n(\mathbf{r}) = \frac{X}{m_\mathrm{p}}\rho(\mathbf{r})$, where X is the cosmic fraction of hydrogen when account is taken of the contribution of other elements i.e. helium.

Suppose now that we switch on a massive star capable of emitting copious amounts of Lyman continuum photons (having energy $h\nu \geq 13.6\,\text{eV}$) and that this position defines the center of a Cartesian co-ordinate system. The photons interact with the neutral medium and produce free electrons (with an assumed number density n_e) and free protons (n_p). Since in our approximation we neglect ionization of helium, we have $n_e = n_p$ (because a hydrogen atom consists of one electron and one proton only).

By invoking the OTS approximation described in Sect. 2.1, the recombination rate per unit volume is $\alpha_B n_e n_p$ and assuming that inside the H II region the hydrogen is fully ionized, $\alpha_B n_e n_p = \frac{\alpha_B \rho^2(\mathbf{r})}{m^2}$, where $m = m_p/X$ and by assuming that $\rho_e(\mathbf{r}) = \rho_p(\mathbf{r}) m_e/m_p$.

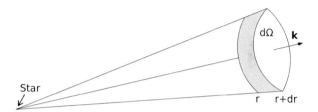

Fig. 2.4 The solid angle $d\Omega$ and the unit vector $\hat{\mathbf{k}}$ emanated from the star (ionizing source)

Consider the element of volume defined by the infinitesimal solid angle $d^2\Omega$ about the unit vector $\hat{\mathbf{k}}$ (as seen from the ionizing source) and the infinitesimal range of radii $(r, r + dr)$, as measured from the star. Figure 2.4 shows this setup. If the number flux of ionizing photons in the direction $\hat{\mathbf{k}}$ is $\dot{N}(r)$, then the equation of ionization balance gives

$$\dot{N}(r)r^2 d^2\Omega = \frac{\alpha_B \rho^2(\mathbf{r})}{m^2} r^2 d^2\Omega\, dr + \dot{N}(r + dr) \cdot (r + dr)^2 d^2\Omega, \qquad (2.5)$$

or if we cancel $d^2\Omega$,

$$\dot{N}(r)r^2 = \frac{\alpha_B \rho^2(\mathbf{r})}{m^2} r^2 dr + \dot{N}(r + dr) \cdot (r + dr)^2. \qquad (2.6)$$

From the above equation we obtain

$$\dot{N}(r + dr) \cdot (r + dr)^2 - \dot{N}(r)r^2 = -\frac{\alpha_B \rho^2(\mathbf{r})}{m^2} r^2 dr. \qquad (2.7)$$

The lefthand side of the latter is

$$\dot{N}(r + dr) \cdot (r + dr)^2 - \dot{N}(r)r^2 = \frac{d}{dr}\left(\dot{N}(r)r^2\right) dr, \qquad (2.8)$$

so if we combine this relation with Eq. 2.7 we obtain

$$\frac{d}{dr}\left(\dot{N}(r)r^2\right)dr = -\frac{\alpha_B \rho^2(\mathbf{r})}{m^2}r^2 dr. \tag{2.9}$$

By canceling dr from both sides and integrating, we obtain

$$\dot{N}(r)r^2 = \frac{\dot{\mathcal{N}}_{\text{LyC}}}{4\pi} - \frac{\alpha_B}{m^2}\int_{r'=0}^{r'=r}\rho^2(\mathbf{r}')r'^2 dr', \tag{2.10}$$

where $\frac{\dot{\mathcal{N}}_{\text{LyC}}}{4\pi}$ is the constant of integration and $\dot{\mathcal{N}}_{\text{LyC}}$ is the total emission rate of the star. Equation 2.10 gives the photoionization balance condition inside an H II region.

Let us consider R_{IF} to be the position of the ionization front (IF), where the flux of the arriving ionizing photons is zero ($\dot{N}_{\text{tot}}(R_{\text{IF}}) = 0$). We define here that the ionization front is the moving edge of the H II region. Assuming that the material inside the H II region is fully ionized, Eq. 2.10 is then

$$\int_{r=0}^{r=R_{\text{IF}}}\rho^2(\mathbf{r})r^2 dr = \frac{\dot{\mathcal{N}}_{\text{LyC}}m^2}{4\pi\alpha_B}. \tag{2.11}$$

If the density distribution is uniform ($\rho(\mathbf{r}) = \rho_0$), then Eq. 2.11 becomes

$$\frac{\rho_0^2 R_{\text{IF}}^3}{3} = \frac{\dot{\mathcal{N}}_{\text{LyC}}m^2}{4\pi\alpha_B}, \tag{2.12}$$

and so the resultant H II region is spherical with radius

$$R_{\text{St}} \equiv R_{\text{IF}} = \left(\frac{3\dot{\mathcal{N}}_{\text{LyC}}m^2}{4\pi\alpha_B\rho_0^2}\right)^{1/3}, \tag{2.13}$$

where R_{St} is the Strömgren radius (Strömgren 1939).

Strömgren (1939) was the first to show that the transition from a state of almost completely ionized material to a state of almost completely neutral material occurs in a very short distance compared to the dimensions of the H II region. Thus we may treat the front as a sharp discontinuity. The distance over which the degree of ionization changes from $\simeq 90\%$ to $\simeq 10\%$ is given by

$$\Delta R_{\text{St}} \simeq \frac{20m}{\rho_0\bar{\sigma}} \simeq 2\times 10^{-4}\text{ pc}\left(\frac{\rho_0}{10^{-20}\text{ g cm}^{-3}}\right)^{-1}. \tag{2.14}$$

Here $\bar{\sigma} = 7\times 10^{-18}$ cm^2 is the photoionization cross section presented by a hydrogen atom to Lyman continuum photons from an OB star.

2.4 Heating and Cooling Mechanisms

As we have previously mentioned, the source of ionizing radiation in H II regions is a single or multiple massive stars, in the spectral type of O or B (hence known as OB stars). An OB star has a typical effective temperature[4] of $T_{eff} \sim 3$–6×10^4 K. Therefore one would expect that such a star would be capable of heating up the surrounding medium in temperatures up to these values. However, typical H II regions have $T_i \simeq 10^4$ K which is much lower than T_{eff}. This difference in temperatures result from various cooling processes that occur simultaneously and that do not allow the ionizing temperature to go higher than $T_i \simeq 10^4$ K.[5] Below we briefly describe the most important heating and cooling sources occurring in an H II region and which result in its thermal balance.

The main heating process in an H II region comes, clearly, from the UV ionizing radiation emitted by the central source. Consider a photon carrying energy more than the 13.6 eV ionization potential of hydrogen, i.e. $h\nu = 15$ eV. When this photon interacts with a hydrogen atom it will ionize it producing a photoelectron. The excess of $15 - 13.6 = 1.4$ eV will be converted to the form of kinetic energy in the photoelectron. In a medium, the faster the particles move, the more hot the medium is, therefore the above kinetic energy simply increases the temperature of the H II region. Other heating processes may apply such as cosmic rays, X-rays, etc. but due to their small contribution in the total heating of the ionized medium we will not discuss them further. If no cooling process was present, the H II region temperature would gradually increase to reach the temperature of the star.

There are several cooling processes that act to decrease the temperature of an H II region down to its observed value. As long as an electron and a proton recombine to form a hydrogen atom, the electron kinetic energy is converted into radiation which can escape the nebula while carrying energy, therefore cooling it. This radiation is able to cool the medium if the recombination does not occur directly to the ground state, otherwise as we have discussed earlier it can ionize a nearby hydrogen atom diffusively. Other cooling processes include the bremsstrahlung radiation which is emitted by the thermal plasma and which is observed at radio frequencies, as well as the collisional excitations of fast moving electrons with neutral hydrogen. These collisions can convert the kinetic energy of an electron to radiation which can also escape the nebula.

However, the most important cooling process is the forbidden line cooling. Species responsible for this effect are metals, such as O and Ne. In case of low abundance in these metals (i.e. low metallicity), forbidden line cooling is not strong enough which act to allow the temperature of the emission nebula to increase.

---

[4]The effective temperature corresponds to the temperature of the object as if it was black body and emitting the equivalent amount of electromagnetic radiation.

[5]In emission nebulae, T_i can reach higher values if we decrease the metallicity, since in this case we change the chemical processes. For the same reason and due to the more powerful heating sources, planetary nebulae and supernova remnants have also in general $T_i > 10^4$ K.

Forbidden lines arise when an electron of a metal is excited by a collision (i.e. with a free fast moving electron) into a metastable state (also known as fine structure levels). These metastable states act as an "energy trap" meaning that when an electron occupies this state, it gets significantly longer time to spontaneously de-excite back to its lower energy value in comparison with the rest of states. For example, a spontaneous de-excitation from a metastable state may take from $t \sim 10$ s to $t > 10^4$ s compared to $t \sim 10^{-8}$ s in the "normal" case. When the medium is dense, during the above period of i.e. 10 s, it is very highly probable for the metal to collide again with another particle of lower energy, hence for the excited electron to transfer its energy excess to the other particle via a collision. Then, the excited electron of the metastable state will lose energy by converting it to kinetic energy. However, if the medium is rarefied enough to avoid these collisions, lower than a value known as "critical density" (n_{crit}), then the electron will spontaneously de-excite to a lower state by releasing a photon. This photon has a frequency which cannot ionize a hydrogen atom, therefore it escapes the nebula contributing in its cooling. These are the known forbidden lines (or fine structure lines) and are very frequently observed in nebulae, particularly in planetary nebulae.[6]

All above heating and cooling mechanisms result in the thermal balance of an H II region which reaches a final temperature of $T_i \simeq 10^4$ K. Forbidden lines can act as nebular diagnostics. They can tell much about the properties of the nebula such as the abundances of species therein, the density of the ionized and atomic medium, the strength and the profile of the ionizing profile of the radiation emitted by the central star and others. These lines are mostly observed in radio wavelengths and (far-)infrared wavelengths, the observation of which requires ground-based arrays such as ALMA and JVLA, or space-borne telescopes such as *Spitzer* and *Herschel* respectively.

2.5 R-Type Expansion

When an ionizing source is switched on in an initially neutral medium (which for simplicity we assume that it is spherically symmetric of uniform number density, n), photons start to rapidly ionize it. The initial phase of ionization is extremely quick with the ionization front reaching velocities significantly higher than the sound

---

[6]Under the same effect the famous 21-cm forbidden line emission of H I is produced however there are two exceptions: (1) the 21-cm line is not emitted from an H II region and does not contribute in its cooling (2) the 21-cm line occurs when there is a spin flip of the electron when compared to the spin of the proton. The more stable case is when the two spins are anti-parallel (para-hydrogen) contrary to the parallel case (ortho-hydrogen). For a spontaneous flip of the ortho to para state, it is required a time of 1–10 million years(!) in order not for the ortho-hydrogen atom to undergo a single collision. It is astonishing to consider that a 21-cm map of a galaxy has been produced by photons that only Astrophysicists can see and that will almost certainly never be reproduced in any lab on Earth.

speed of the ISM. This velocity is so high that the medium cannot hydrodynamically response and it therefore remains approximately undisturbed. This initial phase is known as 'R-type expansion' (R: Rarefied).

If we consider that $\mathcal{N}_{LyC}$ is the number of photons emitted by an ionizing source and assuming that the medium remains undisturbed once the photons start to propagate in it causing ionizations, we can determine the position of the expanding IF as follows. During the R-type expansion phase, photoionization equilibrium is not met hence there will be always a surplus of photons remaining after a given volume has been ionized. This is expressed as

$$\dot{\mathcal{N}}_{surplus} = \dot{\mathcal{N}}_{LyC} - \frac{4}{3}\pi r_{IF}^3 \alpha_B n^2. \tag{2.15}$$

The above equation tells us that the remaining photons are equal to the total number emitted by the source minus those that have been lost by recombinations of electrons with protons to other than the ground state (i.e. OTS approximation). The surplus amount of photons are causing ionizations to an expanding volume equal to $4\pi r_{IF}^2 dr_{IF}/dt$ (that is the surface of the sphere at r_{IF} times the velocity of expansion). This expanding volume has n number density therefore

$$\dot{\mathcal{N}}_{surplus} = n4\pi r_{IF}^2 \frac{dr_{IF}}{dt}. \tag{2.16}$$

Combining Eqs. 2.15 and 2.16 we obtain the differential equation

$$n4\pi r_{IF}^2 \frac{dr_{IF}}{dt} = \dot{\mathcal{N}}_{LyC} - \frac{4}{3}\pi r_{IF}^3 \alpha_B n^2. \tag{2.17}$$

Using the Strømgren radius relation (Eq. 2.13), we solve the above to find the position of the IF in the R-type phase

$$r_{IF}(t) = R_{St} \left(1 - e^{-n\alpha_B t}\right)^{1/3} \tag{2.18}$$

where α_B is the case-B recombination coefficient (see Sect. 2.1) and t is the time. The R-type expansion will be terminated as soon as $e^{-n\alpha_B t} \sim 0$ which occurs when $t_{rec} \sim 3/n\alpha_B$. The quantity $t_{rec} = 1/n\alpha_B$ is commonly referred to as "recombination time". When the R-type phase terminates, the Strømgren sphere has been created and the system has obtain photoionization balance as we have explained above. Thereafter, the dynamical expansion begins which is described in the section below.

2.6   D-Type Expansion

Once the Strømgren sphere has been formed and photoionization equilibrium has been met, the H II region continues to expand due to the high pressure difference observed between the ionized and the neutral medium. This expansion phase, called

'D-type expansion' (D: Dense), continues until pressure equilibrium is met. Once the system obtains this equilibrium, no further dynamical evolution of the H II occurs. In the D-type expansion, the H II region expands at a high-supersonic speed relative to the sound speed of the neutral medium. This drives the formation of a strong shock front ahead of the expanding ionized region (i.e. the ionization front) which sweeps material in time.

> To understand why a shock front is formed, it is important that we should describe what the sound speed does represent. The sound speed is the speed at which a disturbance is propagating in a given medium. The sound itself is actually a disturbance. Consider the atmospheric air at the sea level. The speed of sound at that level is ∼0.3 km/s. If a moving object (i.e. a supersonic airplane) exceeds that speed, it means that the disturbance it creates (from its motion) cannot move faster than the object itself. Hence material of the atmospheric air is constantly being packed up in the area just ahead of the edge of the object towards its moving direction.

For the standards of the ISM, the sound speed, c_s is connected with the gas temperature according to the relation:

$$c_s = \sqrt{\frac{k_B T_{gas}}{\mu m_p}} \tag{2.19}$$

where k_B is Boltzmann's constant and μ is the mean molecular weight of the gas. A medium consisted purely by atomic hydrogen (so $\mu = 1$) at $T_{gas} = 10\,K$ has a sound speed of $c_o \sim 0.2\,km/s$. If this medium is heated up to $T_{gas} = 10^4\,K$ it is ionized (so $\mu = 0.5$) and its sound speed is much higher $c_i \sim 12.8\,km/s$. As we have seen, these temperatures are the standard (average) temperatures found in H II regions. The speed at which an H II region expands exceeds the sound speed of the neutral medium by approximately an order of magnitude, hence a shock front is formed.

The motion of an expanding H II regions associated with the formation of a shock front has been discussed initially by Kahn (1954). However, Spitzer (1978) was the first to attempt an analytical model of the D-type expansion and further discussions followed by Dyson and Williams (1980) have explored this model in more depth. In this approximation, the time-dependent thermal pressure $P_i(t)$ of the ionized gas is $P_i(t) = \rho_i(t)c_i^2$, where c_i is the sound speed of the ionized medium. However, the first to fully explore the D-type expansion to the point of pressure equilibrium was Raga et al. (2012a) while in Raga et al. (2012b) further additions concerning the inertia of the moving shocked material have been considered.

Raga et al. (2012a) considered a model in which the two pressures of (1) the neutral gas in the shell confined between the ionization front and the shock front and of (2) the ram pressure of the undisturbed neutral material (i.e. the one that the shock front has not yet overrun) are equal. They then approximated the D-type expansion of the modelled H II region by using the *piston relation* in which the velocity of the shock front, V_{SF}, is equal to the velocity of the ionization front, $\dot{R}$, plus the post-shock velocity, V_{ps}. However, V_{ps} can be replaced with $V_{SF}/\mathcal{M}_o^2$, where $\mathcal{M}_o = V_{SF}/c_o$ is the Mach number, therefore

$$V_{SF} = \dot{R} + \frac{c_o}{V_{SF}}. \tag{2.20}$$

Under the above assumptions, the velocity of the shock front can be also expressed in terms of the Strømgren radius, the position of the ionization front and the sound speed of the ionized medium as

$$V_{SF} = \left(\frac{R_{St}}{R}\right)^{3/4} c_i, \tag{2.21}$$

and by replacing Eq. 2.21 to 2.20 we obtain the differential equation

$$\frac{1}{c_i}\dot{R} = \left(\frac{R_{St}}{R}\right)^{3/4} - \left(\frac{c_o}{c_i}\right)^2 \left(\frac{R}{R_{St}}\right)^{3/4}. \tag{2.22}$$

Its solution gives the position of the ionization front in unit time. In the above differential equation, the second term on the righthand side can be neglected at early times, since both ratios c_o/c_i and R/R_{St} are in general small. In this case, Eq. 2.22 has the simple solution

$$R(t) = R_{St}\left(1 + \frac{7}{4}\frac{c_i t}{R_{St}}\right)^{4/7}, \tag{2.23}$$

which is known as "Spitzer equation" (Spitzer 1978). However, as the H II expands, the ratio $(R/R_{St})^{3/4}$ becomes much larger than $(c_o/c_i)^2$ and the negative term in the righthand side of Eq. 2.22 starts to dominate. In this case, the H II region stagnates and no further expansion is allowed. This stagnation radius is given by the relation

$$R_{STAG,I} = \left(\frac{c_i}{c_o}\right)^{4/3} R_{St}. \tag{2.24}$$

The expansion of an H II region however, can be also studied using the equation of motion of the expanding shell (note that in the method described above, the differential equation resulted by equating the pressures acting on the shell). This has been studied by Hosokawa and Inutsuka (2006) and independently by Raga et al. (2012b), the latter of which considered additionally the external ram pressure of the undisturbed medium on the H II region. Raga et al. (2012b) described in their work that the differential equation of motion of the shell i.e.

$$\frac{d}{dt}\left(M\dot{R}\right) = 4\pi R^2 \left(P_i - P_o\right) \tag{2.25}$$

includes the inertia of the shocked gas. In particular, if we expand Eq. 2.25, we then end up to a second order differential equation of the form

$$\ddot{R} + \left(\frac{3}{R}\right)\dot{R}^2 = \frac{3R_{St}^{3/2}c_i^2}{R^{5/2}} - \frac{3c_o^2}{R}. \tag{2.26}$$

If we were to omit the $\ddot{R}$ term, Eqs. 2.22 and 2.26 are then the same. Similarly to what we have discussed previously, the term c_o^2/R is in general small at early times when the H II region expands, thus we may neglect it. In this case, Eq. 2.26 has as analytical solution

$$R(t) = R_{St}\left(1 + \frac{7}{4}\sqrt{\frac{4}{3}}\frac{c_i t}{R_{St}}\right)^{4/7}. \tag{2.27}$$

This solution was presented by Hosokawa and Inutsuka (2006). It differs by Eq. 2.23 by the $\sqrt{4/3}$ factor which, according to Raga et al. (2012b), results from the inclusion of the inertia. As the ionized medium expands further, however, the negative term of Eq. 2.26 starts to dominate and, as before, the H II region stagnates. The stagnation radius obtained here is given by the relation

$$R_{STAG,II} = R_{St}\left(\frac{8}{3}\right)^{2/3}\left(\frac{c_i}{c_o}\right)^{4/3} \tag{2.28}$$

2.6.1   Further Discussion

Figure 2.5 plots the solutions of Eq. 2.22 (Raga-I), Eq. 2.26 (Raga-II), Eq. 2.23 (Spitzer) and Eq. 2.27 (Hosokawa-Inutsuka). During two different workshops,[7] many groups focusing on computational star formation emphasizing on expanding H II regions gathered to validate their codes against the above set of equations (Bisbas et al. 2015). The results of their simulations agreed on a different expansion law, marked as "Simulations" in Fig. 2.5. This law follows in the early times of the D-type expansion the Raga-II and Hosokawa-Inutsuka equations, but at later times it stagnates to the value of Raga-I equation, which is the generalized case of Spitzer's approximation. During intermediate times, it is between the Raga-I and Raga-II equations.

---

[7]The STARBENCH workshops in 2013 (Exeter, U.K.) and in 2014 (Bonn, Germany).

Currently, there is no universal equation that can explain this behavior. However, we may provide here some remarks that can help us towards this. Once the H II region starts expanding, a strong shock front forms. At early times, this shock front is quite thin and very dense. As time progresses, the shock front continues to sweep up material from the undisturbed medium, however it tends to expand somehow faster than the ionization front as long as the latter follows a $\propto t^{4/7}$ law. This means that at later times (i.e. when the H II region stagnates) the shock front will be completely detached. Hence, the stagnated H II region will not have a shock front at all immediately after the ionization front.

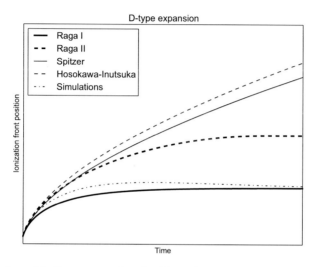

Fig. 2.5 Solutions of the D-type expansion of an H II region. The *dot-dashed line* shows the result of the 12 different codes participating in the STARBENCH project

This points to the assumption that Spitzer's well known equation and its generalized form by Raga et al. (2012a) (Raga-I) is an, in fact, *no-shell* approximation i.e. at all times no shock front is developed. On the contrary, the Hosokawa-Inutsuka equation as well as its generalized form by Raga et al. (2012b) (Raga-II) is produced by solving the equation of motion of the expanding shell. In this way, it *is* considered the development of a shock front which is infinitesimally thin (thin shell approximation), meaning that it has an infinite density (δ-function).

Raga-I and Raga-II are then acting as boundary conditions in which a family of different expansion laws exist. Due to the conditions assumed in the simulations of the STARBENCH project, the codes followed the particular dash-dotted law of Fig. 2.5. However, work in this direction has to been done by the community which will give a much better understanding of how an H II region expands.

References

Osterbrock, D. E. 1974, Research supported by the Research Corp., Wisconsin Alumni Research Foundation, John Simon Guggenheim Memorial Foundation, Institute for Advanced Studies, and National Science Foundation. San Francisco, W. H. Freeman and Co., 1974. 263 p.

Wood, D. O. S., & Churchwell, E. 1989, ApJS, 69, 831

Menten, K. M., Reid, M. J., Forbrich, J., & Brunthaler, A. 2007, A&A, 474, 515

Getman, K. V., Flaccomio, E., Broos, P. S., et al. 2005, ApJS, 160, 319

Preibisch, T., Kim, Y.-C., Favata, F., et al. 2005, ApJS, 160, 401

Bally, J., O'Dell, C. R., & McCaughrean, M. J. 2000, AJ, 119, 2919

Lucas, P. W., & Roche, P. F. 2000, MNRAS, 314, 858

Henry, R. B. C., Kwitter, K. B., & Dufour, R. J. 1999, ApJ, 517, 782

López-Martín, L., Raga, A. C., Mellema, G., Henney, W. J., & Cantó, J. 2001, ApJ, 548, 288

O'dell, C. R., & Handron, K. D. 1996, AJ, 111, 1630

Meaburn, J., Clayton, C. A., Bryce, M., et al. 1998, MNRAS, 294, 201

Burkert, A., & O'Dell, C. R. 1998, ApJ, 503, 792

Matsuura, M., Speck, A. K., McHunu, B. M., et al. 2009, ApJ, 700, 1067

Walch, S. K., Girichidis, P., Naab, T., et al. 2014, arXiv:1412.2749

Mackey, J., Mohamed, S., Gvaramadze, V. V., et al. 2014, Nature, 512, 282

Davidson, K., & Fesen, R. A. 1985, ARA&A, 23, 119

MacAlpine, G. M., & Satterfield, T. J. 2008, AJ, 136, 2152

Hester, J. J., Mori, K., Burrows, D., et al. 2002, ApJ, 577, L49

Lyubarsky, Y. E. 2002, MNRAS, 329, L34

Hester, J. J., Stone, J. M., Scowen, P. A., et al. 1996, ApJ, 456, 225

Barlow, M. J., Swinyard, B. M., Owen, P. J., et al. 2013, Science, 342, 1343

Owen, P. J., & Barlow, M. J. 2015, ApJ, 801, 141

Nutter, D., Ward-Thompson, D., & André, P. 2006, MNRAS, 368, 1833

Stamatellos, D., Whitworth, A. P., & Ward-Thompson, D. 2007, MNRAS, 379, 1390

Strömgren, B. 1939, ApJ, 89, 526

Kahn, F. D. 1954, Bull. Astron. Inst. Netherlands, 12, 187

Spitzer, L. 1978, New York Wiley-Interscience, 1978. 333 p.,

Dyson, J. E., & Williams, D. A. 1980, New York, Halsted Press, 1980. 204 p.,

Raga, A. C., Cantó, J., & Rodríguez, L. F. 2012a, Rev. Mexicana Astron. Astrofis., 48, 149

Raga, A. C., Cantó, J., Rodríguez, L. F., & Velázquez, P. F. 2012b, MNRAS, 424, 2522

Hosokawa, T., & Inutsuka, S.-i. 2006, ApJ, 646, 240

Bisbas, T. G., et al. 2015, MNRAS, 453, 1324

Chapter 3
Triggered Star Formation

Abstract This chapter discusses the most important modes that trigger the formation of stars in expanding H II regions. In Sect. 3.1 we give a brief overview of the observational characteristics that point where and how star formation may be triggered in nebulae. In Sect. 3.2 we describe the very basic physics followed in a gravitationally unstable cloud and in particular we describe the Jeans criterion, Jeans mass and Jeans length that are frequently used in literature. Section 3.3 presents the two dominant hydrodynamical instabilities found in expanding H II regions and which create the potential sites of triggered star formation. These are the Vishniac instability (Sect. 3.3.1) and the Rayleigh-Taylor instability (Sect. 3.3.2). The two sections that follow up (Sects. 3.4 and 3.5 respectively) discuss the Collect and Collapse mechanism and the radiation-driven implosion mechanism for triggered star formation that are observed in expanding H II regions. We close this chapter by discussing how our Solar System was formed (Sect. 3.6) as all observational evidence point to the fact that it was closely connected with a massive star that underwent a supernova explosion.

3.1 Overview

Infrared observations of the Galactic plane from different space observatories such as the *Spitzer Space Observatory* or the *Herschel* space telescope, provide an ubiquitous presence of round-shaped 'bubbles'. Their common characteristic is that their edges are usually more bright than the rest of the surrounding medium indicating that they are more dense. These bubbles can be associated with stellar winds or even supernova explosions. However, they are most likely to be associated with H II regions meaning that in their interior one or more massive stars might exist emitting UV radiation capable of ionizing the surrounding environment. For example, Churchwell et al. (2006, 2007) identified $\sim$600 bubble-shaped objects emitting infrared radiation while Deharveng et al. (2010) concluded that $\sim$85 % of them are associated with H II regions. As we discussed in Chap. 2, an H II region is in overpressure due to the high thermal pressure difference between the hot ($T_i \sim 10^4$ K) and the cold ($T_o \sim 10$ K) temperatures of the ionized and the neutral or molecular regions respectively. This overpressure drives the expansion of

© Springer International Publishing Switzerland 2016
T.G. Bisbas, *The Interstellar Medium, Expanding Nebulae and Triggered Star Formation*, SpringerBriefs in Astronomy, DOI 10.1007/978-3-319-26142-3_3

the H II region creating the dense shells seen in these infrared images. As we will see in this chapter, it is in this shell that star formation is triggered and hence it is interesting to study them both theoretically and observationally.

Classifying the shapes of these bubbles is perhaps a difficult task. First of all we must express that bubbles can be categorized as small-scale (of the order of light years or a few pc) and large-scale (i.e. > 3–4 pc). The small-scale bubbles appear to have a more spherical morphology, whereas the large-scale bubbles have more complex and rather diverse morphologies. The latter is due to the fact that the ISM appears to have irregular structure with dense and rarefied regions making it to be far from uniform. Hence the ionizing photons, although we can well approximate their emission from a potential exciting source as spherically symmetric, propagate at different distances as they interact with different density gradients in each different direction. This in turn makes a large-scale emission nebula to contain bright rimmed clouds (Urquhart et al. 2009) and pillars (Sugitani et al. 2002; Preibisch et al. 2012). However, there are also cases that indeed the surrounding ISM can still be approximately uniform thus making the bubble to appear spherically symmetric, such as the RCW120 nebula (Anderson et al. 2010).

The bubbles are potential sites for triggered star formation. Various theoretical studies confirmed by several observational surveys show that this indeed occurs particularly in the dense expanding shell (in the case of a spherical bubble) or in the dense clumps and tips of pillar structures (in the case of more irregular bubbles). However, the ionizing radiation emitted by massive stars can also result in a negative feedback, that is to suppress the accretion flow in to the collapsing region to form a new star (Dale et al. 2005). It is interesting to note that it is impossible just by observing the star forming region to claim whether or not the stars have been triggered (Dale et al. 2015). However, triggered star formation can not only be restricted to the interaction of ionizing radiation with the ISM. Other mechanisms including cloud-cloud collisions, tidal arcs around interacting galaxies and triggered gravitational instability ranging from spiral-arm dust lanes can also compress the gas to form stars. However in this book, our discussion will be limited to triggered star formation associated with expanding H II regions.

Various high-resolution observations show the existence of small neutral clumps in the interior of an H II region the formation of which is associated with the clumpy and irregular ISM while it is overrun by the supersonic expanding ionization front. Once they are overrun by the ionization front, these pre-existing clumps in the neutral gas start to develop a conic cometary tail pointing away from the ionizing source as a result of shadowing behind the clump. This structure is commonly referred to in the literature as an *elephant trunk* and the dense clump as an *Evaporating Gaseous Globule* (EGG). If it is sufficiently dense, an EGG can be the birthplace of a YSO.

Figure 3.1 shows a picture of the IC1396 nebula, also known as the "Elephant's Trunk" nebula. In this stunning picture, one can observe the different structures formed resulting from the interaction of the ionizing radiation emitted by the exciting source. This exciting source, is the massive star HD 206267 and it is pointed in the middle of the picture. All surrounding material is quite rarefied and

Fig. 3.1 The wider region of the IC1396 nebula. The elephant trunk (enlarged embedded image on *top right*) is observed to contain young stellar objects in its interior. The *other two enlarged areas* correspond to dark nebulae or—due to their size—to Bok globules. The *bright indicated star* in the middle of the image is the massive star HD206267 which is responsible for the UV photons emission. Filters used: Hα, O III, S II. Photo taken at the Machon Observatory Panorama Thessaloniki by Kallias Ioannidis

nearly completely ionized. There are, however, three interesting features. The first one (righthand enlarged image) shows a detail of the elephant trunk. It is easily characterized by its sharp and bright edge. This edge is quite dense and it does not allow photons to penetrate deeper, allowing molecular hydrogen to survive. It is in this innermost place that gravitational collapse may occur (see Sect. 3.2) resulting in star formation. Indeed, many YSOs have been observed in its interior (i.e. Reach et al. 2009) with the youngest to be embedded inside the head of the elephant trunk and the slightly older (but still very new stars) to be embedded in the vicinity of the trunk but inside the H II region. All those are much younger than the central HD 206267 star indicating that they were formed after the H II region has been well developed. This mode is also supported by the simulations of Mackey and Lim (2010) and further by the observations of Patel et al. (2015). Dense material is also observed in other parts in this figure i.e. in the lower enlarged image, which is additionally a potential site for star formation. Smaller features can be also seen (lefthand enlarged image) which are tiny globules surviving the pass of the ionization front. Due to their low mass, these globules may never be able to form a new object and they are simply extremely slowly ionized.

3.2 Gravitational Collapse

In this section we will briefly describe the conditions that need to be satisfied in order for gravitational collapse to begin. The overall physics of star formation, however, is a much more complicated task and outside the scopes of this book. Hence, we will only discuss the simplified case for stability of a spherically symmetric cloud without going through the complicated issues of further internal processes that occur. So when we refer to "star-formation" we shall refer to "collapsing gas" which will inevitably form an object satisfying hydrostatic equilibrium. If its final mass exceeds $\sim 0.08\,M_\odot$, nuclear reactions will be ignited and a new star will form. A GMC in the ISM is frequently subject to various perturbations occurring by different phenomena. For example, a collision of the GMC with another cloud (cloud-cloud collision) or its interaction with a nearby shock front propagating due to a supernova explosion or due to an expanding H II region (such as we explore in this book) do seed perturbations which may change the dynamical state of the parent cloud.

Let us consider a spherically symmetric cloud that is consisted of uniform density, ρ, and has radius, R. Its mass is therefore $M = 4\pi\rho R^3/3$. The cloud is initially relaxed and in isothermal state with temperature T. In order for the cloud to be in equilibrium, the following condition must be satisfied:

$$2K + \Omega = 0 \tag{3.1}$$

where K is its internal kinetic energy and Ω its gravitational potential energy. The above condition is known as the *virial theorem* and if it is met we call the cloud *virialized* or that it is in *virial equilibrium*. Its internal kinetic energy is given by the equation

$$K = \frac{3}{2}Mc_s^2 \tag{3.2}$$

where c_s is the sound speed of the medium at the above given temperature and density. The gravitational potential energy is given by the equation

$$\Omega = -\frac{3}{5}\frac{GM^2}{R} \tag{3.3}$$

By replacing Eqs. 3.2 and 3.3 to Eq. 3.1, we find that the radius that the cloud needs to have in order to be in equilibrium is

$$R_J = \sqrt{\frac{15c_s^2}{4\pi G\rho}} \tag{3.4}$$

The above quantity is referred to as *Jeans length* after Sir James Jeans who was the first to study the conditions leading to gravitational collapse in 1902. If we replace Eq. 3.4 to its mass relation, we obtain

$$M_J \sim 5.5 \frac{c_s^3}{G^{3/2} \rho^{1/2}} \qquad (3.5)$$

If a cloud has radius $R > R_J$ it means that the internal thermal energy is larger than the gravitational potential energy thus the cloud will expand. If $R < R_J$ the opposite effect occurs in which the gravitational potential energy is larger thus the cloud undergoes collapse. Taking into account Eq. 3.5, the collapsing cloud should always have mass $M > M_J$.

Once the cloud collapses, its density increases but the conservation of energy requires gravitational potential energy to be converted to thermal energy. In case the thermal energy (which is in fact radiation) escapes the cloud, there is then sufficient cooling and we may consider that the temperature, T, does not significantly vary (isothermal case of collapsing cloud). Thus the sound speed, c_s, will not change and the Jeans criterion may still be valid and the collapse continues. However, there is eventually the case that the density of the medium increases so high that the medium becomes opaque and radiation is trapped. In this case, the gas is heated up and so the thermal energy is additionally increased. The battle between the thermal energy and the gravitational potential energy may now switch and the collapse phase is possible to be halted and in some cases even reversed (expansion of cloud). Thus for a better understanding of the star formation process, one needs to know the thermodynamics involved as it is of significant importance to determine whether or not a star will be formed or not.

Depending on the initial mass and extend of the unstable GMC, the collapsing phase may result in the formation of one or of several different stable objects (i.e. stars). For example, in a gravitationally unstable cloud (which is usually massive and extended although exceptions may apply), the Jeans condition may be satisfied in several places simultaneously internally to the collapsing region. In this case the cloud *fragments* and if there is sufficient cooling the different fragments may undergo further collapse forming multiple objects simultaneously. This mode is referred to as *hierarchical fragmentation* and the first person to consider it was Hoyle in 1953. With this mechanism, a star cluster is formed such as the open cluster of Pleiades (see Fig. 1.3), or the spherical (globular) cluster of M80 in the Scorpius constellation. Three-dimensional hydrodynamical simulations including self-gravity by Bate et al. (2003), Bonnell et al. (2003) and other researchers have showed how such a cluster may be formed from the gravitational collapse of a GMC. In case there are additional external pressures acting on the collapsing cloud, the fragmentation process may be more violent than before and hence more massive objects may form resulting in extended and compact spherical clusters known as 'super star clusters' which are quite rare. The object Westerlund 2 is an example of a super star cluster which consists of more than ten massive stars. We also note that massive star formation (i.e. the exact process at which the mass of a star reaches

$\gtrsim 25 \, M_\odot$) is quite complicated and thus far unknown to astronomers with several groups worldwide to propose various theoretical models using high resolution and detailed simulations.

3.3 Instabilities During the D-Type Phase

As we discussed above, a star is formed when a dense clump becomes gravitationally unstable. Expanding H II regions may trigger the collapse of pre-existing density enhancements. It interesting to note that some of those dense clumps may not have been able to collapse otherwise, showing that ionizing radiation may result in a positive feedback for star formation. During the D-type expansion of an H II region, there are various hydrodynamical instabilities developed which may result in the formation of several other density enhancements. Here we will focus on the two most important which are the *Vishniac* and the *Rayleigh-Taylor* instabilities.

3.3.1 Vishniac Instability

The *Vishniac* instability (Vishniac 1983) occurs in a decelerating pressure-driven shell and it has been suggested that this instability plays a considerable role in the fragmentation of the ISM. It explains the filamentation of expanding shells as seen by various observations. However, as pointed by Michaut et al. (2012) using hydrodynamical simulations, fragmentation does not always happen as suggested by theory meaning that further investigation needs to be done.

Consider an H II region expanding into a uniform medium which drives the formation of a dense shell (shock front). In Fig. 3.2 we show how this instability acts in this expanding shell. At the bottom, which marks the time $t = 0$, the shell (gray layer) is completely undisturbed and without any perturbation yet formed. Here, the thermal pressure due to the hot H II region, P_i (solid arrows), is perpendicular to the shell. In the outer part of the shell i.e. in the neutral region, the ram pressure, P_o (dashed arrows), is always in radial and opposite to the expansion direction. In this particular case it is perpendicular to the shell, hence no instability forms. During the expansion however (i.e. at $t > 0$) small perturbations occur which make the shock front departing from its stable and uniform shape. As these perturbations grow, the two pressures P_i and P_o are not aligned anymore as shown in the middle layer of Fig. 3.2 and irregularities start to grow.

Eventually, the material inside the shell will start flowing creating locally dense and rarefied regions. Many authors refer to these regions as "hills" and "valleys". Their velocities will be smaller and greater relative to the velocity of propagation of the H II region respectively, creating alternating and oscillating density concentrations and rarefractions. This oscillation is shown in the middle and upper layer of Fig. 3.2. Vishniac (1983) showed that the amplitude of these oscillations grows as $t^{1/2}$. Their separation, which is the wavelength of the instability, is comparable to the thickness of the shell. Observations in expanding ionized regions such as the planetary nebula Retina (IC 4406) or in the Veil nebula (Fig. 3.3) show the existence of structures with the morphologies predicted by the Vishniac instability.

Fig. 3.2 The Vishniac instability. The *bottom panel* shows the shell before any instability grows. Here the thermal pressure (*solid arrows*) is parallel to the ram pressure (*dashed arrows*) of the neutral medium. During the expansion of the shell perturbations may appear which breaks the alignment between the two pressures. In the *middle layer*, the thermal and ram pressures are not always parallel. Note that the thermal pressure is *always* perpendicular to the shell whereas the ram pressure is *always* in radial and opposite to the expansion direction. The flow of material inside the shell makes "hills" and "valleys" which are alternating causing the effect of an oscillating expanding shell as seen in the *upper panel*

To understand better the morphology of an expanding shell in which the Vishniac instability has been developed, take an A4 sized piece of paper. Rumple it well and then unwrap it by pressing it everywhere to obtain its initial A4 size. The unwrapped paper will now have regions which look like hills and valleys. This morphology (in this case anaglyph) is perhaps the closest you may get to such a shell using very simple material.

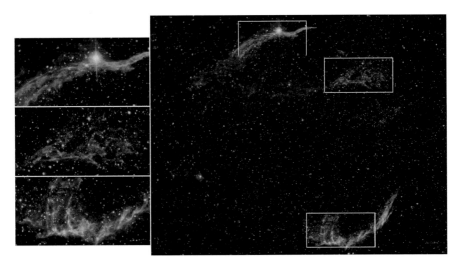

Fig. 3.3 The Veil nebula. The filaments observed are believed to be due to the Vishniac instability. Filters used: Hα, O III, S II. Photo taken at the Machon Observatory Panorama Thessaloniki by Kallias Ioannidis

3.3.2   Rayleigh-Taylor Instability

Of great importance in the astrophysical context is the Rayleigh-Taylor instability. In general, it occurs when a fluid of density n_1 is accelerating a fluid of density n_2 and where $n_1 < n_2$. Small perturbations start to develop which magnify as a function of time. In the astrophysical context, these perturbations result in small (compared to the overall size) and elongated pillar-like structures as seen in expanding H II regions.

Figure 3.4 sketches a sequence of the development of a Rayleigh-Taylor instability in an ionized medium. The vectors inside the shell illustrate how the material is flowing towards the elongated pillar structure. The region around the axis of symmetry of this pillar is denser as material flows. Note that in this figure $d < h$ indicating that the rest of the shell has been further expanded than the denser tip at the same time. Hence the elongated middle part becomes thinner in time. If it becomes thin enough, the spheroidal tip can be completely detached thus floating inside the H II region. Once this happens, these tiny globules are exposed in an approximately uniform thermal pressure acting from all sides which may then act to change its internal density structure. Eventually after some several sound crossing times the final density profile is likely to have a smooth shape. Such tiny globules dubbed as "globulettes" have been observed by Gahm et al. (2007) in the Rosetta Nebula. Figure 3.5 shows a detailed image of the Rosetta nebula in which the morphological structures described here are present. It is interesting to note that those globulettes are the result of the passage of the ionization front while it was expanding. They are not the result of any other process that may have occurred inside the ionizing medium.

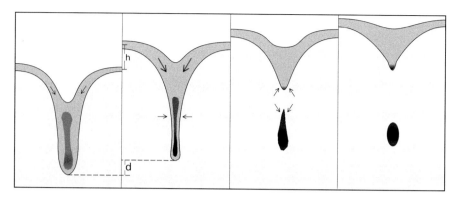

Fig. 3.4 This sketch shows the sequence followed during the Rayleigh-Taylor instability. The ionizing radiation is impinging from *bottom* to *top*. The *grayscale* corresponds to density (the *darker* it is, the more dense it is). The shock front here is thought to expand in vacuum—hence there is *plain white color* ahead of it—in order to achieve an accelerating expansion (otherwise the expansion is $\sim t^{4/7}$). If a perturbation occurs, the shock front is prone to the Rayleigh-Taylor instability and a finger-like structure (i.e. a pillar) is formed as shown above (time increases from *left to right*). Inside this finger-like structure, neutral material is flowing creating a dense head which may be detached later on and survive the strong UV radiation while embedded in the H II region. This can result in a proplyd (dense small globules frequently observed inside the ionized medium). It is interesting to note that such globules are a result of the passage of the ionization front. If dense enough, they may gravitationally collapse to form new stars or even free floating brown dwarfs

Note that these dense clumps surviving inside the hot environment of an H II region need not always result due to the Rayleigh-Taylor instability. Their morphological sequence, however, is very reminiscent to what is described here. For example, early work by Kahn (1958) showed that the elephant trunks observed in H II regions cannot be formed through instabilities in planar-ionization fronts. In order for the Rayleigh-Taylor instability to appear in an expanding H II region, one of the following two general cases are needed to be fulfilled: (a) either the material ahead the shock front to be more rarefied than the density of the ionized medium in the interior of the H II region or (b) the ionizing source to rapidly increase by orders of magnitude its radiation output such as when a supernova explosion occurs. A great example for the second case is the elongated pillars appearing in the Crab Nebula (Hester et al. 1996). In both cases the lighter (more rarefied) material pushes and accelerates the heavier (denser) region. Williams (1999) has further investigated instabilities in the ionization front using hydrodynamical simulations that will lead in star formation in the dense neutral clumps within an H II region.

The Rayleigh-Taylor instability can be reproduced in a simple experiment using a glass of cold water in which one carefully adds a layer of food-colored

(continued)

warm water. As the warm water cools down and due to the addition of the color it becomes slightly denser than the water below it and hence finger-like structures start to appear which have small spheroids in their tips. Here, gravity acts to "accelerate" (by pulling downwards) the heavier food-colored water.

Fig. 3.5 Detail of the Rosetta nebula. The *dark structures* are dense material which took this shape due to hydrodynamical instabilities developed during the expansion of the H II region. The *left enlarged image* shows a pillar while the *right one* a small globulette embedded in the ionized region (compare with Fig. 3.4). Filters used: Hα, O III, S II. Photo taken at the Machon Observatory Panorama Thessaloniki by Kallias Ioannidis

3.4   Collect and Collapse

The *collect-and-collapse* (C&C) mode for triggering star formation has been first proposed by Elmegreen and Lada (1977) and has been studied in greater detail by Whitworth et al. (1994). This mode is believed to be a simple mechanism leading to star formation triggered by an expanding H II region. In order to understand how the C&C mechanism works, let us assume the case of a spherically symmetric cloud of uniform density in which we switch on an ionizing source in its center. The evolution of the newly formed H II region has been explained in Sect. 2.5 and in Sect. 2.6. As the spherical shock front expands, it sweeps up neutral material from the undisturbed medium that overruns. Hence, the mass of the shell increases while at the same time its temperature can be considered to stay approximately constant. This is the first 'C' of the C&C mechanism ('Collect' phase) in which material is constantly packed up at the shock front.

As the shock front expands further, its mass, M_{shell}, in places becomes comparable to the Jeans mass, M_J, and is therefore prone to gravitational instability. This self-gravitating attraction corresponds to the second 'C' of the C&C mechanism ('Collapse' phase) and can lead to the formation of gravitationally bound objects (fragments). If massive enough (i.e. exceeding Jeans' mass), these fragments collapse further possibly followed by star formation.

You may compare the collect-and-collapse mechanism that possibly triggers star formation by considering the example in which a large group of friends are about to meet in an elongated table and discuss various subjects. Suppose that in that table, not everyone arrives at the same time but some friends arrive earlier than others (beginning of the expansion phase of an H II region). These begin some discussion about a random subject. As time progresses (shock front development), more and more friends arrive and try to participate in this discussion (collect phase). Although this can be achievable for some time, there will be a point at which, after some critical number of friends, N_{crit}, a second discussion by someone will begin simultaneously attracting the interest of the nearby friends (collapse phase). Therefore, the initial subject has been unstable against the increasing number of friends and has therefore fragmented. The more the number of friends, the more the number of different subjects are expected to be discussed (further collect and collapse phase as the H II region keeps expanding). Each arriving friend represents the new material that is swept up by the shock front as it expands. Each different subject of discussion represents a different fragment.

The C&C mechanism has been observed in many expanding H II regions. In RCW79, Zavagno et al. (2006) report evidence for massive fragments at the shock front of the expanding H II region in which twelve luminous new stars have been identified. In addition, Zavagno et al. (2007) report eight different fragments in the RCW120 nebula, five of which are located at the borders of the expanding H II region, with the most massive of them to be $\sim$370 $M_\odot$. As in RCW79, young stellar objects have been also identified. In both those nebulae, the age and the position of the observed new protostars point that the C&C mechanism is responsible for their formation. Other groups have also reported similar observations in various other objects. The C&C mechanism has been also modelled with three-dimensional hydrodynamical codes (i.e. Dale et al. 2007; Walch et al. 2011). The results show how this process of triggered star formation occurs in time thus offering a better understanding of this mechanism.

3.5 Radiation-Driven Implosion

Another important mechanism responsible for triggered star formation in expanding H II regions is *radiation-driven implosion* (RDI). The term was first introduced by Sandford et al. (1982) to describe the effect in which neutral clumps are exposed in external ionizing radiation.[1] As we saw in Chap. 2, pre-existing clumps in the neutral gas may result in the formation of elephant trunks and EGGs which are potential sites of star formation. It is interesting to examine the evolution of neutral clumps inside the H II region which were primarily formed due to the overrun of the ionization front from the clumpy ISM. RDI occurs when a clump or small cloud interacts with UV radiation emitted by an ionizing source which is placed outside and far away from it (Bertoldi 1989; Lefloch and Lazareff 1994; Kessel-Deynet and Burkert 2003; Haworth and Harries 2012).

The RDI mechanism and its connection to triggered star formation is an area that both observational and theoretical groups are trying to understand. In particular, Deharveng et al. (2005) addresses the main questions underlying this connections as to "*where* (in the core, or at its periphery) and *when* (during the maximum compression phase, or earlier) star formation takes place". In an attempt to answer these questions, high resolution simulations presented by Bisbas et al. (2011) showed that the location and the star formation time (indicating when the first star has been formed) depends on (1) the initial mass of the clump and (2) the intensity of the ionizing radiation.

In particular, that work suggests that low fluxes of ionizing radiation favor star formation increasing its efficiency. In this case, the overall dynamical evolution of the clump is slow when compared to the star formation timescale. The new stars are found to be formed towards the inner part of the clump (i.e. some distance ahead of the ionization front). These new stars are likely to have relatively high mass, as the accretion phase timescale is sufficient enough to support such a mass growth. The density of this small region inside the clump where star formation takes is high which thus extinguishes the ionizing radiation to that depth. Eventually, the ionization front propagates further around this dense region which results in the formation of a cometary tail. These objects are known as *bright-rimmed clouds* (BRCs) and are of great observational interest. A collection of BRCs can be found in the SFO catalog (Sugitani et al. 1991; Sugitani and Ogura 1994). Some of these BRCs host YSOs in their tip indicating that star formation has been triggered.

---

[1] However, there are some authors who describe this effect as *radiation-driven compression*.

3.6 The Formation of Our Solar System

Various groups worldwide have suggested that our solar system has been formed in the vicinity of a massive star that underwent supernova explosion. It is not clear however if the shock wave of the explosion has triggered the collapse of the molecular cloud to form the protoplanetary disk, or if the collapse would inevitably have occurred. The observational evidence that supports the above formation scenario comes from the short lived radioactive isotopes (known as SLRs or SLRI) identified in meteorites. An SLR is an element that has the same number of protons as the respective stable elements, but with different number of neutrons. Such SLRs are elements like ^{26}Al, ^{41}Ca, ^{53}Mn, and ^{60}Fe. The particular ^{60}Fe isotope is of great importance as it is perhaps an excellent indicator of a nearby supernova explosion (Busso et al. 1999).

^{60}Fe has a half-life of $\tau \simeq 2.6$ Myr and is made only in the late stages of a massive star before the explosion. To produce ^{60}Fe, it is required a high flux of neutrons and therefore sufficient amount of this SLR can only be formed in massive stars. The ejecta of the supernova explosion contains ^{60}Fe that has not yet decayed, known as "live" ^{60}Fe. Live ^{60}Fe has been observed in the early solar system (Shukolyukov and Lugmair 1992) providing very strong evidence for the formation of our solar system near a massive star. SLRs can also be used as chronometers for the formation time of the protoplanetary disk. For example, as shown by Jacobsen et al. (2008), the decay of ^{26}Al to ^{26}Mg and which has $\tau \simeq 0.7$ Myr can be considered as the highest resolution clock for investigating the solar system formation time.

Hester and Desch (2005) proposed a six-stage scenario that supports the mode of triggered collapse in a supernova environment that can explain the solar system formation. Those six stages are the following: (1) compression of molecular gas around the edge of an H II region (i.e. either the C&C or RDI mechanisms), (2) gravitational collapse of this molecular gas and formation of a $\sim 1 M_\odot$ star, (3) the formation of a cometary tail object such as a BRC containing an EGG in its tip, (4) formation of a protoplanetary disk the surrounding shell of which is exposed to ionizing radiation forming a proplyd, (5) a phase in which this proplyd survives embedded in the interior of the H II region since its rarefied surroundings have been 'cleared' by the interaction of the ionizing radiation, and (6) a supernova explosion of the nearby massive star the ejecta of which enrich the medium and the environment of the proplyd with SLRs. Parts of the above sequence can be seen in various nebulae such as NGC 281 (Pacman Nebula; see Fig. 3.6).

The above stages have been reproduced with high resolution hydrodynamical simulations. Gritschneder et al. (2012) using two-dimensional simulations examined supernova shock waves interacting with an initially spherically symmetric cloud. They found that it is possible to enrich a pre-existing clump sufficiently with the SLR ^{26}Al to explain the observed ratio in the Solar system and they claim that this clump should have been ~ 5 pc away from the massive star. Boss and Keiser (2012) performed the first ever fully three-dimensional simulations that examine triggered star formation in supernova explosions. They found that Rayleigh-Taylor instability is responsible for the finger-like structures observed in such nebulae and that it is in these places that star formation is observed. The simulated abundance enrichment of ^{60}Fe is in agreement with the observed one of our Solar system. In addition Boss and Keiser (2014) have further concluded that the massive star probably underwent at Type II supernova explosion.

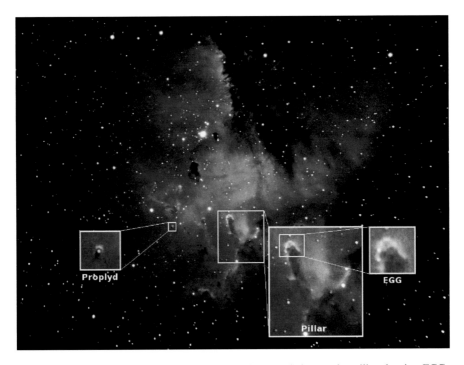

Fig. 3.6 NGC 281 (The Pacman nebula). This emission nebula contains pillars hosting EGGs at their tips and which are dense enough to gravitationally collapse and form new stars. As the ionization front expands, it 'clears' the region around the pillar leaving the small dense clump embedded in the H II region thus creating a proplyd. It is believed that our Solar system may have followed the above sequence. Soon after the supernova explosion of the nearby massive star, the environment was enriched with heavy elements and SLRs. Filters used: Hα, O III, S II. Photo taken at the Machon Observatory Panorama Thessaloniki by Kallias Ioannidis

What about the overall environment of the collapsing cloud that formed our Solar system? Using N-body methods for investigating they dynamics and gravitational interactions, Pfalzner (2013) examined the idea of the formation of the Solar system in the environment of a star cluster. She found that in this case, the cluster is highly probably to have been in an OB association. OB associations are loose star clusters containing few tens of massive stars of spectral type O and B and hundreds to thousands of lower mass stars (such as our Sun). This finding can also explain the circularity observed in the Solar system.

Nevertheless, all observational evidence we currently have point to the fact that our presence is due to a supernova explosion occurred about 4.6 billion years ago, followed by a sequence of events which happened by the sheerest accident at the right time to create the conditions appropriate for life to form. And if this happened to us, it will have been happened somewhere else as well. Therefore, we may conclude that triggered star formation and the likelihood of life may be much closer related than we currently think.

References

Churchwell, E., Povich, M. S., Allen, D., et al. 2006, ApJ, 649, 759
Churchwell, E., Watson, D. F., Povich, M. S., et al. 2007, ApJ, 670, 428
Deharveng, L., Schuller, F., Anderson, L. D., et al. 2010, A&A, 523, A6
Urquhart, J. S., Morgan, L. K., & Thompson, M. A. 2009, A&A, 497, 789
Sugitani, K., Tamura, M., Nakajima, Y., et al. 2002, ApJ, 565, L25
Preibisch, T., Roccatagliata, V., Gaczkowski, B., & Ratzka, T. 2012, A&A, 541, A132
Anderson, L. D., Zavagno, A., Rodón, J. A., et al. 2010, A&A, 518, L99
Dale, J. E., Bonnell, I. A., Clarke, C. J., & Bate, M. R. 2005, MNRAS, 358, 291
Dale, J. E., Haworth, T. J., & Bressert, E. 2015, MNRAS, 450, 1199
Reach, W. T., Faied, D., Rho, J., et al. 2009, ApJ, 690, 683
Mackey, J., & Lim, A. J. 2010, MNRAS, 403, 714
Patel, N. A., Sicilia-Aguilar, A., & Goldsmith, P. 2015, American Astronomical Society Meeting Abstracts, 225, #211.03
Bate, M. R., Bonnell, I. A., & Bromm, V. 2003, MNRAS, 339, 577
Bonnell, I. A., Bate, M. R., & Vine, S. G. 2003, MNRAS, 343, 413
Vishniac, E. T. 1983, ApJ, 274, 152
Michaut, C., Cavet, C., Bouquet, S. E., Roy, F., & Nguyen, H. C. 2012, ApJ, 759, 78
Gahm, G. F., Grenman, T., Fredriksson, S., & Kristen, H. 2007, AJ, 133, 1795
Kahn, F. D. 1958, Reviews of Modern Physics, 30, 1058
Hester, J. J., Stone, J. M., Scowen, P. A., et al. 1996, ApJ, 456, 225
Williams, R. J. R. 1999, MNRAS, 310, 789
Elmegreen, B. G., & Lada, C. J. 1977, ApJ, 214, 725
Whitworth, A. P., Bhattal, A. S., Chapman, S. J., Disney, M. J., & Turner, J. A. 1994, A&A, 290, 421
Zavagno, A., Deharveng, L., Comerón, F., et al. 2006, A&A, 446, 171
Zavagno, A., Pomarès, M., Deharveng, L., et al. 2007, A&A, 472, 835
Dale, J. E., Bonnell, I. A., & Whitworth, A. P. 2007, MNRAS, 375, 1291
Walch, S., Whitworth, A., Bisbas, T., Hubber, D. A., & Wuensch, R. 2011, arXiv:1109.3478
Sandford, M. T., II, Whitaker, R. W., & Klein, R. I. 1982, ApJ, 260, 183
Bertoldi, F. 1989, ApJ, 346, 735
Lefloch, B., & Lazareff, B. 1994, A&A, 289, 559
Kessel-Deynet, O., & Burkert, A. 2003, MNRAS, 338, 545
Haworth, T. J., & Harries, T. J. 2012, MNRAS, 420, 562
Deharveng, L., Zavagno, A., & Caplan, J. 2005, A&A, 433, 565
Bisbas, T. G., Wünsch, R., Whitworth, A. P., Hubber, D. A., & Walch, S. 2011, ApJ, 736, 142
Sugitani, K., Fukui, Y., & Ogura, K. 1991, ApJS, 77, 59
Sugitani, K., & Ogura, K. 1994, ApJS, 92, 163
Busso, M., Gallino, R., & Wasserburg, G. J. 1999, ARA&A, 37, 239
Shukolyukov, A., & Lugmair, G. W. 1992, Lunar and Planetary Science Conference, 23, 1295

Jacobsen, B., Yin, Q.-z., Moynier, F., et al. 2008, Earth and Planetary Science Letters, 272, 353
Hester, J. J., & Desch, S. J. 2005, Chondrites and the Protoplanetary Disk, 341, 107
Gritschneder, M., Lin, D. N. C., Murray, S. D., Yin, Q.-Z., & Gong, M.-N. 2012, ApJ, 745, 22
Boss, A. P., & Keiser, S. A. 2012, ApJ, 756, L9
Boss, A. P., & Keiser, S. A. 2014, ApJ, 788, 20
Pfalzner, S. 2013, A&A, 549, A82

Chapter 4
Numerical Simulations

Abstract This chapter discusses the importance of numerical simulations and presents some basic applications of expanding H II regions. After a brief introduction (Sect. 4.1) we give a description of the most important methods that the computational community follows (Sect. 4.2). These include grid-based methods (Sect. 4.2.1), smoothed particle hydrodynamics methods (Sect. 4.2.2) and hybrid methods (Sect. 4.2.3) which are new mesh-moving techniques coupling the advantages of the two previous methods. In Sect. 4.3 we present applications of a spherically symmetric expanding H II region (Sect. 4.3.1), an H II region expanding off-centered in a spherical cloud (Sect. 4.3.2), a neutral clump exposed to external radiation impinging from one side (Sect. 4.3.3) and an H II region expanding in a fractal cloud (Sect. 4.3.4) in which all features previously discussed are present. We close the chapter (Sect. 4.4) by describing the importance of synthetic observations and how these can be helpful towards our understanding of the observational data cubes.

4.1 Introduction

The core of scientific method is the validation of theoretical predictions using experiments. In natural sciences, the experiments can usually take place in a lab. For the science of Astrophysics however, this is perhaps an impossible scenario as the conditions examined cannot be reproduced, at least with the current technology. For example, the lowest density ever achieved in a lab is of the order of $\gtrsim 10^3 \, \text{cm}^{-3}$ which is much larger than the density typically found in the ISM. This density is already much lower (by many orders of magnitude) than the density of $\sim 10^{13} \, \text{cm}^{-3}$ found at $\sim 100 \, \text{km}$ above the surface of the Earth and which defines the boundary of space.[1] Numerical simulations, therefore, stand as the lab for Astrophysicists and hence detailed developments in this regard are the only possible way to

---

[1] On the contrary, very low temperatures—even lower than the temperature of the cosmic microwave background radiation—have been reproduced. The 2001 Nobel Prize in Physics has been awarded to Eric A. Cornell, Wolfgang Ketterle, and Carl E. Wieman for this achievement.

© Springer International Publishing Switzerland 2016   51
T.G. Bisbas, *The Interstellar Medium, Expanding Nebulae and Triggered Star Formation*, SpringerBriefs in Astronomy, DOI 10.1007/978-3-319-26142-3_4

understand and explain observational data, as long as pure theoretical and analytical investigations are quite limited and in most cases quite simplified to explain complex phenomena.

Many groups worldwide attempt to tackle each individual problem in Astrophysics using numerical simulations. In these numerical simulations, the user adopts the appropriate equations and chemical reactions needed in order to examine the problem in question. The results of such calculations can be then compared to observations and predictions arising directly from the corresponding theoretical study. Numerical simulations are recently considered as the new pillar for scientific method. In 2013, the Nobel Prize in Chemistry has been given to three researchers[2] "for the development of multiscale models for complex chemical systems". That was the first Nobel Prize ever to be given to numerical simulations. In parallel, the great advancement in the technology of large supercomputers has allowed the community to explore three-dimensional systems including very complex Physics and large chemical networks constituted by many thousands of reactions.

Taking into account the evolution of numerical simulations in the community of Astrophysics studying Star Formation and the ISM during the past two decades or so, one can divide them in two different kinds of categories. The first one corresponds to the study of (hydro-)dynamical evolution including gravity, magnetic fields and advanced thermodynamics. The second one corresponds to the study of Astrochemistry which examines the evolution of chemical reactions in a static density distribution (i.e. a system which does not dynamically evolve) under ISM conditions. During the period that this book was written, the community is rather focused on constructing extended codes that couple dynamical evolution with Astrochemistry. This will ultimately allow us to a much deeper understanding of the life cycle of the ISM inside our Galaxy or in distant objects (i.e. other galaxies even during early Epoch times), how stars and planets are formed and how at the same time chemical species evolve and are distributed in the Galactic and Extragalactic environment.

There are two basic methods that community follows to validate a numerical code. The first one is to reproduce a known problem—usually simplified—which has been already analytically solved prior to the test and then compare the results of the experiment (simulation) with the corresponding theoretical predictions. Although this is a standard and widely acceptable methodology, the increase of the complexity of numerical codes requires the test against more complicated problems which cannot be solved analytically. In this case, the second validating method is the one that individual codes simulate exactly the same problem in question, the results of which are then being compared against each other. This validating method is known as 'benchmarking' and is more accurate with increasing the number of participating codes.

---

[2]Martin Karplus, Michael Levitt, Arieh Warshel.

In the following, we will fully concentrate on dynamical simulations and we will explore basic cases of expanding nebulae pointing to how star formation is triggered, after a brief discussion on the methods followed by the majority of community.

4.2   Methods Followed in Numerical Simulations

In Star Formation and in Astrophysics in general, the evolution of the interstellar gas is examined using the fluid equations. Consider a simple case of a fluid in which we neglect magnetic fields and self-gravity. The corresponding equations are then as follows. The *continuity equation* is

$$\frac{d\rho(\mathbf{r})}{dt} = -\rho(\mathbf{r})\nabla \cdot \mathbf{v}(\mathbf{r}) \tag{4.1}$$

where $\rho(\mathbf{r})$ is the density of the fluid and $\mathbf{v}(\mathbf{r})$ its velocity at position $\mathbf{r}$. The above equation describes the conservation of mass. The derivative d/dt is the Lagrangian time derivative, i.e. the rate of change experienced by an observer moving with the fluid:

$$\frac{d}{dt} = \frac{\partial}{\partial t} + \mathbf{v} \cdot \nabla . \tag{4.2}$$

The *momentum equation* is

$$\frac{d\mathbf{v}(\mathbf{r})}{dt} = -\frac{\nabla P(\mathbf{r})}{\rho(\mathbf{r})} + \mathbf{a}_{grav}(\mathbf{r}) + \mathbf{a}_{visc}(\mathbf{r}), \tag{4.3}$$

where $P(\mathbf{r})$ is the pressure at position $\mathbf{r}$, $\mathbf{a}_{grav}(\mathbf{r})$ is the gravitational acceleration and $\mathbf{a}_{visc}(\mathbf{r})$ is the viscous acceleration. The first term of the righthand side is the hydrostatic acceleration. The *energy equation* is

$$\frac{du(\mathbf{r})}{dt} = -\frac{P(\mathbf{r})}{\rho(\mathbf{r})}\nabla \cdot \mathbf{v}(\mathbf{r}) + \frac{\Gamma - \Lambda}{\rho(\mathbf{r})} + \mathcal{H}_{visc}, \tag{4.4}$$

where $u(\mathbf{r})$ is the thermal energy per unit mass at position $\mathbf{r}$, Γ and Λ are respectively the heating and cooling rates per unit mass due to different additional functions and $\mathcal{H}_{visc}$ is the heating rate due to viscous forces.

By implementing a hydrodynamical algorithm to solve these equations, we are able to simulate the evolution of a given fluid. Many groups worldwide have constructed such codes and almost all of them are based on two different methodologies, namely 'Grid-based' and 'Smoothed Particle Hydrodynamics'. During the last few years or so however, there are other methods developed based on moving meshes and using subdivisions of the computational domain by adopting the Voronoi tessellation and the Delaunay triangulation (the latter of which is related to the Voronoi tessels). A short review of the computational methods used in Astrophysics can be seen in Hubber (2014).

4.2.1 Grid-Based Methods

In grid-based methods (or "finite-volume"), the computational domain is considered to be a large 'box' containing the object (fluid) that we want to simulate. This large computational volume is divided into points or 'cells' and each cell occupies a fraction of the initial volume. Assuming that the domain is subdivided by a uniform set of cells (grid), its resolution (which is a power of 2) is simply defined by their number along all N spatial dimensions. For example, if $2^8 = 256$ is the resolution per direction then 256^N cells is the resolution of the entire simulation. Such methods were first presented almost three decades ago by Colella and Woodward (1984), Roe (1986), and Berger and Colella (1989). These methods are frequently used by many groups worldwide.

Grid-based algorithms are solving the *advection equation*, which in the simplified 1D case, it has the form

$$\frac{\partial q}{\partial t} + u \frac{\partial q}{\partial x} = 0. \tag{4.5}$$

Here, $q = q(x, t)$ is a time-dependent function of the spatial dimension x, and u a constant velocity in time and space. In order to solve the advection equation, one needs to consider two streams: the *upstream* (or *upwind*) which indicates the direction at which the flow comes from, and the *downstream* (or *downwind*) which indicates the direction at which the flow goes. In order to numerically solve the advection equation, we need to approximate Eq. 4.5 as follows:

$$\frac{q(x_2, t_2) - q(x_2, t_1)}{t_2 - t_1} + u \frac{q(x_2, t_1) - q(x_1, t_1)}{x_2 - x_1} = 0, \; u > 0. \tag{4.6}$$

The left panel of Fig. 4.1 shows a 2D cell which has size $\Delta x = \Delta y = d$, therefore it covers an area of d^2. We denote with F the stream along the x axis and with G the stream along the y axis. The center of the cell has coordinates x and y. The two downstream flows in this case are F_1 and G_1, and the two upstream flows are F_2 and G_2 for the x- and y-axes respectively. Nevertheless, the problem of solving the advection equation as accurately as possible is quite complicated and outside the scopes of this book.

Significant efforts have been made also in constructing additional methods for increasing the resolution of a hydrodynamical simulation while keeping the computational cost low. In such demanding computations, a simulation running on 128 or more CPU cores constantly for weeks is quite common. On the other hand if we increase the number of cells in a hydrodynamical run from i.e. $(2^8)^3$ to $(2^9)^3$, it means that we may require 8 times (or more) further calculations and which immediately implies that it becomes computationally expensive. One of the common techniques to solve this problem, is the development of algorithms known as "Adaptive Mesh Refinement". In these algorithms, a cell is subdivided in further cells if a user-defined condition is fulfilled. Take for example the density criterion: in case the density of a cell $\rho(x)$ is $\rho(x) > \rho_{\text{critical}}$, then we need to increase *locally* the resolution in order to simulate further this specific region. At the same time, the neighboring cells may keep the resolution they initially had thus spending approximately the same computational time. A cell will be subdivided until reaching a user-defined maximum resolution.

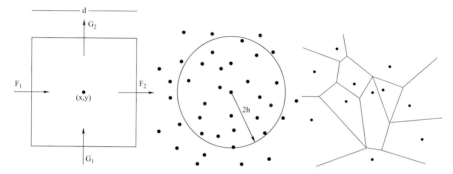

Fig. 4.1 The *left panel* shows a 2D grid of size d which is centered in the (x, y) coordinates. F_1 and G_1 are the *downstreams* and F_2 and G_2 the *upstreams* respectively. The *middle panel* shows SPH particles distributed in the domain. Each particle is considered to have a spherical extend of radius $2h$ in which a number of neighboring particles exist. The *right panel* shows a 2D Voronoi tessellation for a few randomly distributed particles. This scheme may use the advantages of both grid-based and SPH-based techniques

4.2.2 Smoothed Particle Hydrodynamics Methods

Smoothed Particle Hydrodynamics (SPH) is a Lagrangian numerical method. It describes a given fluid by dividing it into N elements which are usually called 'particles'. The higher the N number is, the better the resolution. Depending on the simulation and the computational power available to the user, N can typically be $\sim 10^6$–10^7 although higher numbers of particles are also being frequently used. SPH was invented independently by Lucy (1977) and Gingold and Monaghan (1977) to simulate hydrodynamical problems in Astrophysics concerning non-axisymmetric and self-gravitating setups. Each SPH particle is considered to cover a spherical volume, the extend of which is known as 'smoothing length' and is usually noted as h. With a sphere of *radius* $2h$, there usually exists a constant number of neighboring

SPH particles. This smoothing length h describes the extend of a kernel function used to smooth out the properties of the fluid the most common of which is density. SPH does not, therefore, use a grid as described in Sect. 4.2.1, and the computational domain can be infinite. The middle panel of Fig. 4.1 shows an SPH distribution and the extend of a smoothing sphere.

The entire SPH method is based on the equation

$$f(\mathbf{r}) = \int \delta(|\mathbf{r} - \mathbf{r}'|)f(\mathbf{r}')d^3\mathbf{r}', \tag{4.7}$$

where f is a function describing a quantity (fluid property) at position $\mathbf{r}$ and δ is Dirac's delta function. If instead of the δ function we consider a kernel function W satisfying the properties that (1) it is normalized i.e.

$$\int W(|\mathbf{r} - \mathbf{r}'|, h)d^3\mathbf{r}' = 1 \tag{4.8}$$

and (2) that as $h \to 0$, $W \to \delta$ i.e.

$$\lim_{h \to 0} W(|\mathbf{r} - \mathbf{r}'|, h) = \delta(|\mathbf{r} - \mathbf{r}'|), \tag{4.9}$$

we may then write Eq. 4.7 as

$$f(\mathbf{r}) = \lim_{h \to 0} \int W(|\mathbf{r} - \mathbf{r}'|, h)f(\mathbf{r}')d^3\mathbf{r}', \tag{4.10}$$

and assuming that h is small we get the smoothed value of the quantity f

$$\langle f(\mathbf{r}) \rangle = \int W(|\mathbf{r} - \mathbf{r}'|, h)f(\mathbf{r}')d^3\mathbf{r}'. \tag{4.11}$$

By inserting the density $\rho(\mathbf{r})$ in the latter equation, we may re-write it as

$$\langle f(\mathbf{r}) \rangle = \int W(|\mathbf{r} - \mathbf{r}'|, h)\frac{f(\mathbf{r}')}{\rho(\mathbf{r}')}\rho(\mathbf{r}')d^3\mathbf{r}'. \tag{4.12}$$

By replacing the integral with a summation over the N SPH particles (Monaghan 1992) and by considering that $\rho(\mathbf{r}')d^3\mathbf{r}'$ represents a mass element, the above equation takes the form

$$\langle f(\mathbf{r}) \rangle = \sum_{j=1}^{N} m_j \frac{f_j}{\rho_j} W(|\mathbf{r} - \mathbf{r_j'}|, h). \tag{4.13}$$

Here the particle j has mass m_j, position $\mathbf{r_j}$ and density ρ_j. As N increases (i.e. by increasing the resolution), h decreases and the above summation converges.

Smoothed Particle Hydrodynamics is frequently used in examining the hydro-dynamical motion of the ISM and the associated star formation. However the community challenges this method in simulations containing large temperature, pressure and density gradients (Agertz et al. 2007). Further developments (Price 2008; Hopkins 2013) have addressed these issues and have provided routines and new methodology to increase the validity of SPH.

4.2.3 Hybrid Methods

Both of the grid-based and SPH-based methods have advantages and disadvantages and the community is debated as to whether or not each method is more accurate than the other. A new class of methods has been recently developed which uses characteristics from both methods described above. SPH is known to experience problems i.e. when simulating strong shock fronts or two-phase media (media which have two different temperatures and are in contact; contact discontinuity). On the other hand, grid-based techniques have limitations when bulk velocities are present and are also known to suffer along the coordinate planes. To overcome these, a new formulation based on the Voronoi tessellation scheme is now used by an increasing number of groups. Springel (2010) was perhaps the first to have implemented a full 3D code based on such a technique. Their methods are frequently used in Cosmological applications. Similarly, Hopkins (2015) also implemented a similar technique by merging the advantages of SPH and Grid-based methods.

The Voronoi tessellation (named after Gregory Voronoy who invented the method in the first decade of 1900) divides a given area consisting of randomly distributed particles in non-overlapping tessels. Neighboring tessels have common boundaries. The area covering each tessel is chosen so as to cover the minimum possible surface (in 2D) or volume (in 3D). The right panel of Fig. 4.1 shows a 2D Voronoi tessellation scheme. If we are to connect the centroid of each Voronoi tessel, we construct then the Delaunay triangulation (or tessellation). Thus the Delaunay triangulation is topologically well connected with the Voronoi tessellation. While such schemes have certain advantages compared to grid-based and SPH methods, their downside is that they are computationally expensive as well as that they are characterized by the high level of complexity that their implementation is demanding.

4.3 Simulations

Below we provide a set of simulations where we consider in all cases a GMC interacting with ionizing radiation emitted by a single massive star. Each different simulation uses a different setup in terms of (1) the position of the star and (2) the density distribution of the material constituting the GMC. The numerical scheme we used is SPH and the code we have used to perform these simulations is SEREN (Hubber et al. 2011; Bisbas et al. 2009).[3]

The ionizing source is considered to be a star emitting UV photons carrying energy $h\nu = 13.6\,\mathrm{eV}$ at a rate of $\dot{\mathcal{N}}_{\mathrm{LyC}} = 10^{49}\,\mathrm{s}^{-1}$. This assumption corresponds to a star with spectral type known as OB. OB stars are in general massive stars with mass $M \sim 30\,M_\odot$ and which are short-lived ($\sim 10\,\mathrm{Myr}$ or less before they undergo supernova explosion). In the simulations below, only the D-type expansion is considered (see Sect. 2.6). The effective temperature of such a star is $T_{\mathrm{eff}} \simeq 3 \times 10^4\,\mathrm{K}$. Such stars can be found in emission nebulae as discussed in Chap. 2.

The GMCs we adopt here have two different types of density distribution. The first one is a uniform density distribution in which the density is the same everywhere in the cloud (with a small exception at the very outer 'skin' of the cloud where the density is somehow lower; however due to its thickness it can be assumed to have the same density too). The second type we assume, is a fractal[4] density distribution. Each fractal cloud is described by its fractal dimension, d_f. In three-dimensions that we consider here, $d_f \simeq 3$ means that the cloud is approximately uniform (which means that the uniform density cloud we adopt here is also a fractal with $d_f = 3$) and if $d_f \sim 2$, the cloud is very filamentary. For the simulations presented here we use $d_f = 2.2$ (more clumpy) and $d_f = 2.8$ (less clumpy).

The mass of the GMC considered here is $M_{\mathrm{GMC}} = 10^4\,M_\odot$ and its extend if $R = 5.2\,\mathrm{pc}$ typically found in such objects. In uniform density distribution, the corresponding number density is $n_\mathrm{H} \sim 10^3\,\mathrm{cm}^{-3}$ whereas in fractal clouds this density (although on average it is the same) may reach $n_\mathrm{H} \simeq 6 \times 10^3\,\mathrm{cm}^{-3}$ or more. Hence in the latter cases the interaction of the GMC with the ionizing source can result in a much different evolution as we will see in Sect. 4.3.4.

The equation of state we use, i.e. the dependency of gas temperature on the density of the medium, is assumed to be of two temperatures only, one for the ionized medium and one for the neutral medium. The gas temperature of the ionized medium is taken to be $T_\mathrm{i} = 10^4\,\mathrm{K}$ and for the neutral medium is taken to be $T_\mathrm{o} = 10\,\mathrm{K}$. These temperatures are commonly found in emission nebulae, hence it is an acceptable approximation. The inclusion of different equation of states is possible, however here we want to demonstrate the most simplified examples in order to address the key differences between those. Such examples in which the physics considered is extremely simplified to explore specific effects are called 'control runs' and are of significant importance to test and validate theoretical models in each case.

---

[3]Statement: the choice of this scheme (Smoothed Particle Hydrodynamics) and the respective code (SEREN) were made due to the accessibility of the present author to these utilities. It is *not* intended to conclude that SPH is superior to any other available numerical schemes developed by the community.

[4]A general description of a fractal cloud is that there exists a repeating pattern at every possible scale—from the smallest possible to the largest including the entire size of the cloud.

In all simulations we invoke the on-the-spot approximation (see Sect. 2.1) and we use a recombination co-efficient into excited states only equal to $\alpha_B = 2.7 \times 10^{-13} \, cm^3 \, s^{-1}$.

4.3.1 Spherically Symmetric Expansion

In this simulation we adopt the uniform density case; this is the simplest possible density distribution we can assume, hence this will remark the starting point in exploring the present set of simulations. In this simulation, the GMC is of uniform density and the star is placed in its center. Once the simulation begins, the Strømgren sphere forms (see discussion in Sect. 2.3) which has an extend of $R_{St} = 0.7 \, pc$ corresponding to $\sim 13.5 \, \%$ of the total GMC size. While at the beginning of the simulation the density inside and outside the Strømgren sphere is the same (since during the R-type expansion phase the gas does not hydrodynamically move) and due to the large temperature difference between the two media (ionized and neutral), a difference in thermal pressures of approximately 4 orders of magnitude consequently occurs. This drives the expansion of the D-type phase and in which a dense shock front is formed sweeping up material from the undisturbed medium while it overruns it.

The four left large panels of Fig. 4.2 show four different stages during this spherically symmetric expansion of the H II region. The figure is color-coded according to the total column density (i.e. the total H-nucleus number density). The brighter the color is, the more dense it is. The four right elongated panels show the corresponding cross-section plots (slices along $x = 0 \, pc$ axis) for the same snapshots but are focused on the internal structure of the expanding dense shock front. It can be seen that the shock front follows an oscillating motion as described in Sect. 3.3.1 meaning that it is prone to the Vishniac instability. This instability may be responsible for the (rather fine) 'granulation' observed in the left four panels and in particular at $t = 0.3$ and $0.4 \, Myr$.

4.3.2 Off-Center Expansion

In this simulation, we consider the properties of the cloud described in Sect. 4.3.1, but we now transfer the ionizing source from the center of the cloud to $R_\star = 3.12 \, pc$ along the $-x$ direction, corresponding to about 60 % of GMC's extend. Figure 4.3 shows a sequence of snapshots taken at four different times. Here, it can be seen (at $t = 0.75 \, Myr$) that the left part of the GMC (with respect to the position of the ionizing star) is completely ionized soon after the start of the simulation and the ionizing material streams away freely in the vacuum. Since the ionizing medium escapes rapidly from this side of the cloud, the thermal pressure of the region interior to the parabolic shock front is consequently decreasing, therefore the expansion

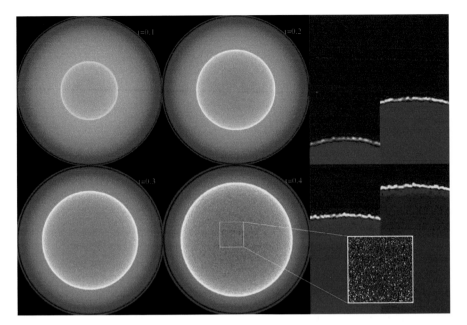

Fig. 4.2 The *four left large panels* show four different stages at times $t = 0.1, 0.2, 0.3$ and 0.4 Myr of a spherically symmetric expanding H II region. The *color* corresponds to the total H-nucleus column density (the *brighter* it is, the more dense it is). The *four elongated panels* on the *right* show the corresponding cross-section plots focusing on the oscillating density structure of the shock front being prone to the Vishniac instability (see Sect. 3.3.1)

from the right side of the ionizing star (i.e. towards the rest undisturbed GMC) is decelerating further than $\propto t^{4/7}$ per unit time. As the shock front evolves in the righthand side, it is initially prone to the Vishniac instability which results in the formation of a perturbed density profile in the interior of the front. At later times when the shock front hits the edge of the cloud (at $t \sim 1.5$ Myr), it starts to expand freely in the vacuum. Since there is no other medium further outside the edge of the GMC to act a ram (opposite) pressure, the shock front expands in an accelerating manner ($t > 1.5$ Myr).

As we discussed in Sect. 3.3.2, the Rayleigh-Taylor instability occurs when a rarefied medium pushes and accelerates a denser medium. This is actually what is happening from this point onwards: the rarefied ionizing medium acts a very high thermal pressure on the much denser shock front and the latter constantly increases its speed per unit time. The ionizing radiation is able to penetrate deeper in the less dense areas of the non-uniform shock front, since they have less column density along the line of sight connecting the front with the ionizing source. This effect of the breakdown of the shock front starts gradually to appear from the outermost parts of its parabolic shape (i.e. those that are most distant from the ionizing star), while towards later times ($t \gtrsim$ Myr) the shell has been entirely fragmented forming small dense globules.

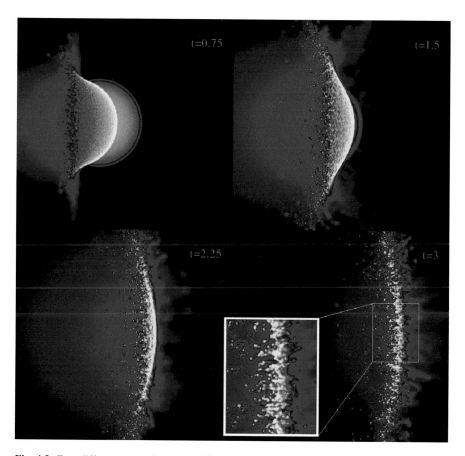

Fig. 4.3 Four different stages during the off-centre expansion of an H II region. The star is placed close to the edge of the GMC. At $t = 0.75$ Myr the H II region has completely ionized the leftmost part of the cloud and the hot material streams away decreasing the overall thermal pressure. At $t = 1.5$ Myr the outermost parts of the shock front start to break forming hundreds of small globules as a result of the Rayleigh-Taylor instability. The shocked shell continues to fragment further ($t = 2.25$ Myr) and at $t = 3$ Myr thousands of small cometary knots have been developed. These are quite reminiscent to the knots observed in the Helix nebula. They have a dense head which may lead to the formation of YSOs

These globules undergo further evolution and behind them a cometary tail forms. Such structures are observed usually in places in expanding nebulae and in particular in planetary nebulae with the Helix one to contain many of such objects. This further evolution of an isolated globule interacting with an ionizing source placed externally and far away from it is the known from Chap. 2 radiation driven implosion, which we will discuss in the next section.

4.3.3   Radiation Driven Implosion

To better understand how a neutral clump or a globule evolves dynamically under the interaction of an external ionizing radiation field, we will consider the following simplified example. Let us assume that the clump is initially represented by a spherically symmetric object of uniform density. We also consider that there is an ionizing source (for example a massive star) placed outside and far away from the clump. Thus we can approximate the impinging radiation as plane parallel. This setup is the one that (Lefloch and Lazareff 1994) considered and provided with two-dimensional simulations which have been reproduced in three-dimensions by Gritschneder et al. (2010), Bisbas et al. (2011), and Haworth and Harries (2012).

Figure 4.4 shows four panels of the evolved clump. The underline dynamical evolution is that the south part of the clump interacting with the ionizing radiation, starts to boil-off and the material in its skin is ionized. This drives a shock front towards the interior of the clump which converges towards its axis of symmetry defined by the line connecting the centre of the clump and the ionizing source. Once the material of the shock front converges from all sides to this axis, it creates an internal 'fountain' of dense material which looks like a small pillar (at $t = 0.12\,\text{Myr}$). It is in this pillar that the highest density of the medium can be found and which under some certain conditions gravitational instabilities may occur thus forming stars. As the clump evolves further (at $t = 0.18\,\text{Myr}$), it creates an elongated cometary tail behind it 'pointing' towards the ionizing source (at $t = 0.24\,\text{Myr}$). The newly formed object is a bright-rimmed cloud as we have also explained in Sect. 3.5.

4.3.4   Fractal Clouds

The ISM density structure is observed to be far from uniform. Many groups worldwide have tried to simulate its spatial distribution as realistically as possible. There have been studies concerning that this observed structure may be a result of gravitational amplification of pre-existing density fluctuations, or from turbulence due to the interactions of ISM with supernovae, outflows and other similar sources. All of those studies have not yet put in place a convincing answer to which one (or which kind of combination) is the dominant factor responsible for the ISM spatial structure.

In order to mathematically construct a cloud containing already some reasonably defined density structure, some groups have adopted the fractal morphology the level of irregularity of which can be easily controlled. Some recent studies of expanding H II regions within fractal clouds have been done by Walch et al. (2012, 2013, 2015) and have applied their outcomes in observed and well known objects such as the RCW 120 object.

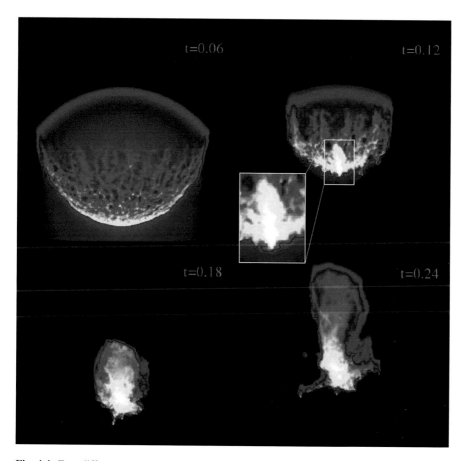

Fig. 4.4 Four different stages during the radiation driven implosion mechanism. The radiation is impinging from *bottom* to *top*, meaning that we consider a star placed outside and far away from the clump. At early times ($t = 0.06$ Myr) a shock front is forming compressing the south part of the clump. The density undulations observed are a result of the Vishniac instability. The material is converging towards its axis of symmetry (i.e. the axis connecting the ionizing source with the centre of the clump) and a dense filamentary structure is formed ($t = 0.12$ Myr). At $t = 0.18$ Myr the sides of the clump have been compressed; it is at this moment that new stars may form. However, due to the internal thermal pressure of the neutral gas, the dense material may bounce back and re-expand, forming the cometary tail ($t = 0.24$ Myr)

As we have discussed in Chap. 3, there are two important modes of triggered star formation: Collect and Collapse and Radiation Driven Implosion. There are objects that only one of the two appears, and objects that both of those modes exist in different places when the nebula is quite extended. The simulations of Walch et al. (2012, 2013, 2015) have shown that in extended emission nebulae i.e. of the order of few parsecs across, the transition between those two modes of triggered star formation depends on the fractal dimension controlling the morphology of the cloud.

For low fractal dimensions the initial cloud has large scale morphological (and hence density) fluctuations, whereas for high fractal dimensions it has smaller scale fluctuations. When a low fractal medium interacts with an ionizing source placed well inside it (i.e. in its centre), the expanding H II region creates in general large patches of shocked material. On the other hand, an expanding H II region in a high fractal medium results in the formation of numerous small globules which in general may have quite high density.

Walch et al. (2015) reported that the two modes of triggered star formation co-exist regardless the morphological structure of the parent GMC, however the low-fractal case results in a *shell-dominated* object (and thus more prone to the Collect and Collapse mechanism) whereas the high-fractal case results in a *pillar-dominated* object (and thus more prone to the Radiation Driven Implosion mechanism).

In Fig. 4.5 we show a simulation of a fractal medium (case of a low fractal GMC) interacting with the ionizing radiation of a star that is placed in its centre. It can be seen that as the H II expands, there are areas which result in the formation of large scale fragments and hundreds of smaller scale globules such as globulettes as we described in Sect. 3.3.2. Recent theoretical work by Haworth et al. (2015) has suggested that globulettes may be the birthplace of free-floating brown dwarfs.

4.4   Synthetic Observations

During the last decade or so, a significant effort towards a better understanding of the observed quantities of the Galactic or extragalactic ISM has been made. Apart from the advance in technology concerning telescopes and astronomical instruments in general, numerical simulations start to play a very important role in this direction. As we have noted in the beginning of this chapter, computational Astrophysics can be basically divided in two different areas: (1) in modelling the dynamical evolution and (2) in modelling the astrochemical evolution of a system (constituting the science of Computational Astrochemistry). In the first area it is mainly assumed that the thermodynamics and thus the equation of state is quite simplified and the gas temperature follows a set of defined functions. Thus a hydrodynamical code does not spend time in calculating this thermodynamical part and hence it is able to focus on solving the equations of motion. In the second area, numerical codes are entirely focused on producing an accurate equation of state by solving a complicated and quite extended network of chemical reactions. By solving these reactions, we are actually solving the very atomic physics under conditions found in space (i.e. low density, high radiation, interaction with cosmic rays etc.) and hence we are able to calculate the emission of different lines that originate from some of the most important species. These emission lines can be directly compared to the observational data sets from radio- and infrared-telescopes, while hydrodynamical simulations can tell much about the evolution of the medium and the star formation process therein.

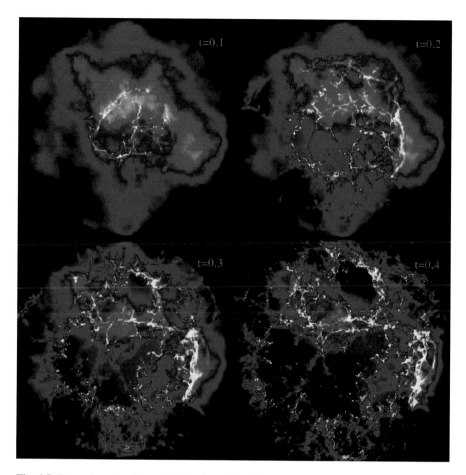

Fig. 4.5 Expansion of an H II region in a fractal cloud. The GMC contains already density clumps before the H II starts to expand. This results in the formation of filamentary structures and local density condensations even at early times ($t = 0.1$ Myr). During the expansion of the ionized medium (i.e. $t > 0.2$ Myr), both collect and collapse and radiation driven implosion mechanisms can be observed. In particular in the *east side* of the cloud, an extended dense shell propagates which may become gravitationally unstable to form YSOs. The *south* and *west parts* of the cloud are in principle dominated by isolated neutral clumps undergoing radiation driven implosion. Note that the *southeast* part of the cloud at $t = 0.3$ Myr is reminiscent to the structure observed in Rosetta nebula (cf. Fig. 3.5)

> Computational Astrochemistry serves as the "bridge" connecting the simulated with the observed data cubes.

To this date, several astrochemical codes have been developed, other examining the ionizing region (i.e. Ercolano et al. 2003, 2005; Ferland et al. 1998), or PDRs

(Röllig et al. 2007) or molecular regions (e.g. Viti and Williams 1999). Further developments and advancements in full three-dimensions are available (i.e. Bisbas et al. 2012). These three-dimensional advancements and the merge of all different chemistry (from ionized and atomic down to molecular) in a single code is the next step in Computational Astrochemistry.

The computational techniques and resources that we currently have, make the implementation of a unified code that includes all dynamical (hydrodynamics, self-gravity, magnetic fields etc.) and chemical (ionized, atomic, molecular) processes in one grand-code a very demanding and extremely difficult task. To understand this, let us consider the following example. If we run a dynamical simulation using a large number of CPUs (i.e. 512) for a total of 5-days (120 h, corresponding to $512 \times 120 = 61,440$ CPU hours), and each timestep requires 60 s for the iteration process (that is the time to solve the equations of motion including all additional complicated physical processes) we will then have a total of 7200 timesteps. An astrochemical simulation considers a static density distribution (i.e. which does not move in time) and solves the chemical network therein. A full 3D simulation may take, under the same number of CPUs, up to i.e. 3 days (72 h or 36,864 CPU hours). Those $\sim 3.6 \times 10^4$ CPU hours serve for only *one* dynamical timestep meaning that a full code that solves dynamics and chemistry simultaneously may take a prohibitively long time even with twice the number of CPUs. Therefore we need to develop new techniques that are able to solve the system of equations much faster than we already do and reduce the computational time by many orders of magnitude.

Currently the community performs synthetic observations in the form of post-process. For instance, detailed dynamical simulations (excluding chemistry) are performed and then individual frames (density distributions) are inserted to astro-chemical codes as initial conditions. Although this is a widely acceptable process, solving the chemistry simultaneously with a dynamical simulation provides a highly detailed equation of state for each particular point which overall affects the thermal pressures acting in every computational element constituting the simulation.

By solving the complete set of chemodynamical equations, we can have very useful information to explore the ISM conditions from small scale objects (i.e. a prestellar core or the formation of a planet) to a whole of extragalactic objects in the Early and distant Universe and understand its properties, spatial density distribution, and star formation process therein.

Acknowledgements I would like to thank Dr. Thomas Haworth for the useful discussions regarding the computational methods followed in astrophysical hydrodynamics.

References

Hubber, D. A. 2014, Astrophysics and Space Science Proceedings, 36, 95
Colella, P., & Woodward, P. R. 1984, Journal of Computational Physics, 54, 174
Roe, P. L. 1986, Annual Review of Fluid Mechanics, 18, 337

Berger, M. J., & Colella, P. 1989, Journal of Computational Physics, 82, 64
Lucy, L. B. 1977, AJ, 82, 1013
Gingold, R. A., & Monaghan, J. J. 1977, MNRAS, 181, 375
Monaghan, J. J. 1992, ARA&A, 30, 543
Agertz, O., Moore, B., Stadel, J., et al. 2007, MNRAS, 380, 963
Price, D. J. 2008, Journal of Computational Physics, 227, 10040
Hopkins, P. F. 2013, MNRAS, 428, 2840
Springel, V. 2010, MNRAS, 401, 791
Hopkins, P. F. 2015, MNRAS, 450, 53
Hubber, D. A., Batty, C. P., McLeod, A., & Whitworth, A. P. 2011, A&A, 529, A27
Bisbas, T. G., Wünsch, R., Whitworth, A. P., & Hubber, D. A. 2009, A&A, 497, 649
Lefloch, B., & Lazareff, B. 1994, A&A, 289, 559
Gritschneder, M., Burkert, A., Naab, T., & Walch, S. 2010, ApJ, 723, 971
Bisbas, T. G., Wünsch, R., Whitworth, A. P., Hubber, D. A., & Walch, S. 2011, ApJ, 736, 142
Haworth, T. J., & Harries, T. J. 2012, MNRAS, 420, 562
Walch, S. K., Whitworth, A. P., Bisbas, T., Wünsch, R., & Hubber, D. 2012, MNRAS, 427, 625
Walch, S., Whitworth, A. P., Bisbas, T. G., Wünsch, R., & Hubber, D. A. 2013, MNRAS, 435, 917
Walch, S., Whitworth, A. P., Bisbas, T. G., Hubber, D. A., Wünsch, R. 2015, MNRAS, 452, 2794
Haworth, T. J., Facchini, S., & Clarke, C. J. 2015, MNRAS, 446, 1098
Ercolano, B., Barlow, M. J., Storey, P. J., & Liu, X.-W. 2003, MNRAS, 340, 1136
Ercolano, B., Barlow, M. J., & Storey, P. J. 2005, MNRAS, 362, 1038
Ferland, G. J., Korista, K. T., Verner, D. A., et al. 1998, PASP, 110, 761
Röllig, M., Abel, N. P., Bell, T., et al. 2007, A&A, 467, 187
Viti, S., & Williams, D. A. 1999, MNRAS, 310, 517
Bisbas, T. G., Bell, T. A., Viti, S., Yates, J., & Barlow, M. J. 2012, MNRAS, 427, 2100

Chapter 5
Summary

In this book we have attempted to present a basic introduction to the interstellar medium, expanding nebulae and triggered star formation in H II regions. A brief introduction to Computational Astrophysics that examines the above aspects including applications has been also included. During the entire book we have attempted to give simplified examples using Earth standards in terms of mass and distance in order the reader to better understand the various difficult concepts presented and that the author tried to address in this work.

The total baryonic mass corresponds to ~4 % of the total composition of the Universe with the rest 96 % to be dark matter (27 %) and dark energy (69 %). In our Galaxy, the ISM corresponds to 75 % of its baryonic mass with the rest 25 % to be stars. Ninety-nine percent of the ISM is in gas phase while the rest 1 % is dust. Yet dust plays an important role controlling the thermal balance of the entire ISM. The average temperature of the ISM (in its molecular form) is ~10–20 K and its average density is $1 \, \mathrm{cm}^{-3}$ (a 500 ml glass full of ISM should only contain 500 particles). The average metallicity of our Galaxy (i.e. proportion of all elements heavier than He in comparison with H I) is one solar metallicity (chemical composition of our Sun) and in general in extragalactic objects metallicity decreases. It is interesting to note again that ~92 % of our bodies is a result of nucleosynthesis occurred in stellar cores and supernova explosions, while the rest ~8 % is a direct product of Big Bang nucleosynthesis. Observing H_2 in our Galaxy and in other galaxies is possible using CO as a tracer, although this method starts nowadays to be challenged. Perhaps cosmic rays (which are actually relativistic particle and *not* rays in terms of radiation) may play an important role in the commonly used method of using CO. An alternative and apparently more powerful tracer than we previously thought is C I.

The ionizing radiation emitted by massive stars changes the chemistry (and hence the physics and dynamics) of their surrounding medium. This UV radiation destroys the atom of H I by producing a free proton and a free electron, while creating the well known nebulae and which are frequently photographed by amateur astronomers

© Springer International Publishing Switzerland 2016
T.G. Bisbas, *The Interstellar Medium, Expanding Nebulae and Triggered Star Formation*, SpringerBriefs in Astronomy, DOI 10.1007/978-3-319-26142-3_5

worldwide (see Fig. 5.1). Nebulae are divided in three categories according to their characteristics: (1) diffuse nebulae (emission, reflection and dark), (2) planetary nebulae which are the end product of low mass stars (including our Sun) and (3) supernova remnants which are the catastrophic explosion of a massive star. The temperature of the H II region of an emission nebula is $\sim 10^4$ K due to the balance of heating mechanisms (ionizing radiation) with cooling mechanisms (such as forbidden line cooling). An H II region is characterized by its ionization front (where the temperature abruptly drops from $\sim 10^4$ K down to ~ 100 K or less). Ahead of the ionization front there is the shock front which results from the supersonic expansion (relative to the sound speed of the neutral medium) of the H II region. The chemistry there changes and from a nearly completely ionized medium we find the atomic medium creating photodissociation regions. PDRs are very important for our understanding of the properties of the ISM as the lines emitted from these regions can tell much about the density, UV radiation field and temperature, and are therefore excellent nebular diagnostics.

Once a massive star is switched on in a spherically symmetric neutral cloud, a very highly supersonic ionization front is propagated from the star to the medium which ends up in forming a sphere (known as the Strømgren sphere, see Eq. 2.13) containing ionized material (R-type expansion phase). However, the high thermal pressure differences between the ionized and the neutral undisturbed medium causes the newly formed H II to expand (D-type expansion phase). There are two different expansion laws that have been proposed. The one resulting by equating the pressures acting on two regions (Eq. 2.22) with a simplified form by Spitzer which is the most well known relation in this aspect (Eq. 2.23) and the second resulting by deriving the equation of motion of the expanding shell (Eq. 2.26 with Eq. 2.27 to be its simplified form). A further study needs to be done in the D-type phase since numerical codes seem to give a somehow different expansion law.

It is well known that ISM is observed to be quite irregular and to contain clumpy clouds. Once an H II expands, the ionization front and the shock front formed are prone to different hydrodynamical instabilities. The two most well known are the Vishniac instability (thin shell instability) and the Rayleigh-Taylor instability. These two act to create density enhancements which are then further compressed by the ionization front. This creates features in a nebula such as bright-rimmed clouds, pillars and other elongated and irregular structures. There usually exist, at the tip of pillars, evaporated gaseous globules (EGGs) which may act as the birth place of young stellar objects (YSOs). A star will occur when the medium becomes gravitationally unstable and collapses under the Jeans criterion (Eq. 3.4). In other words, the expansion of an H II may trigger the formation of new stars. It is believed that our own Solar system may be a result of such a process as chemical analysis in meteorites suggest that the observed short lived radioactive isotopes (SLRs) such as ^{60}Fe may only be a result of a massive star explosion. Two important modes for triggered star formation exist: (1) the collect and collapse mode (which occurs when the mass of the shock front as it expands has grown to such values that fragmentation occurs and which may lead to the formation of stars) and (2) the radiation driven implosion mode (which occurs when a neutral clump is being ionized by a source

Fig. 5.1 The California emission nebula (NGC 1499) in the Perseus constellation. Filters used: Hα, O III, S II. Photo taken at the Machon Observatory Panorama Thessaloniki by Kallias Ioannidis

placed outside and far away from it). There is currently a suggestion that the fractal dimension of a giant molecular cloud which interacts with an ionizing source may act as the critical condition to switch from one mode to the other.

Numerical simulations serve as the lab of Astrophysicists and Astrochemists as space conditions cannot be achieved by manmade machines. Algorithms have started to develop nearly three decades ago or more, however only the last one or two decades have presented three-dimensional and high resolution simulations which are able to investigate from how the dynamics of the ISM and how it evolves, to the formation of stars and, even further, to the formation of disks, brown dwarfs and planets. Numerical simulations can be considered as an additional (and rather new) pillar of scientific method. Although computational astrophysics and computational astrochemistry have both (but separately) developed in great detail, new efforts are being made nowadays to merge those two large areas and develop new, unified, highly complicated and computational demanding codes. With such codes in hand (which need to pass several individual sanity tests before their usage) it will be allowed to the astrophysical community to compare numerical directly with observational data cubes thus making a significant step for a *much* deeper understanding of the physics and chemistry of the ISM. It can be therefore said that the present and perhaps the next decade may consist the golden era of computational astrophysics. A full lifecycle of the ISM, from gas to stars and back to gas again, including microphysics and with resolution down to low mass objects is perhaps the holy grail of Computational Astrophysics.

Printed in the United States
By Bookmasters